ON MUSCLE

ALSO BY BONNIE TSUI

Why We Swim

American Chinatown

FOR CHILDREN

Sarah and the Big Wave

ON MUSCLE

The Stuff That Moves Us and Why It Matters

Bonnie Tsui

ALGONQUIN BOOKS
OF CHAPEL HILL
2025

Copyright © 2025 by Bonnie Tsui. All rights reserved.

Hachette Book Group supports the right to free expression and the value of copyright. The purpose of copyright is to encourage writers and artists to produce the creative works that enrich our culture.

The scanning, uploading, and distribution of this book without permission is a theft of the author's intellectual property. If you would like permission to use material from the book (other than for review purposes), please contact permissions@hbgusa.com. Thank you for your support of the author's rights.

Algonquin Books of Chapel Hill / Little, Brown and Company
Hachette Book Group
1290 Avenue of the Americas, New York, NY 10104
algonquinbooks.com

First Edition: April 2025

Algonquin Books of Chapel Hill is an imprint of Little, Brown and Company, a division of Hachette Book Group, Inc. The Algonquin Books name and logo are trademarks of Hachette Book Group, Inc.

The publisher is not responsible for websites (or their content) that are not owned by the publisher.

The Hachette Speakers Bureau provides a wide range of authors for speaking events. To find out more, go to hachettespeakersbureau.com or email hachettespeakers@hbgusa.com.

Little, Brown and Company books may be purchased in bulk for business, educational, or promotional use. For information, please contact your local bookseller or the Hachette Book Group Special Markets Department at special.markets@hbgusa.com.

ISBN 978-1-64375-308-9 (hardcover)
ISBN 978-1-64375-729-2 (e-book)
Cataloging-in-publication data for this title is available from the Library of Congress.

Printing 1, 2025
LSC-C

Printed in the United States of America

For Andy, who knows what strength is

CONTENTS

Introduction   1

STRENGTH

1. What's Power, in a Body?   9
2. Muscle as Potential   22
3. A Heavy Lift   31
4. The Making of a Hero, Then and Now   39

FORM

5. The Ideal Body   51
6. Who's Afraid of a Lady Hercules?   61
7. Shoulders, Squared   71

ACTION

8. Your Muscles Are Talking   91
9. Jumpology   99
10. The Movement Is the Message   115

FLEXIBILITY

11. Muscles, Fast and Slow — 123
12. It Comes from Unity — 139
13. Remembrance of Exercises Past — 152

ENDURANCE

14. What We Carry — 171
15. Running to Remember — 176
16. True Grit — 185
17. Going the Distance — 190
18. Kissing the Ground in Equilibrium — 203

Epilogue — 209
Acknowledgments — 215
Notes — 219

ON MUSCLE

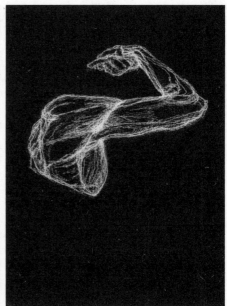

Deltoid, triceps, and biceps brachii (facing away).
Illustration by the author

INTRODUCTION

M ake me a muscle.
At five, six, seven, eight years old, I knew to stick my arm out obligingly and contract my biceps. My father, passing through the room on his way somewhere else, would give my upper arm a squeeze and laugh. "Very good," he'd say.

Then he'd make a muscle back and ask, "Am I fit or what?" It became a family joke.

My father, who moved from Hong Kong to New York in the late 1960s, was more an acolyte of Bruce Lee than of Jack LaLanne. But he'd long been an attentive multidisciplinary student of what I'll call Muscle Academy. Everything from practicing judo, tae kwon do, and karate—the latter two in which he earned a brown belt and a black belt—to steeping himself in fitness Americana: bodybuilding competitions on TV, magazine subscriptions to *Muscle & Fitness*, sketches of famous athletes. He was a professional artist who, among many other accomplishments, created the posters advertising the 1984 Olympic Games on ABC, and, with them, the glorification of the competitors: our modern gods on Earth.

We always had a makeshift home gym, equipped with a motley

collection of free weights, hand grips, and pull-up bars, as well as nunchucks, jump ropes, and heavy punching bags. As far back as memory serves, my brother and I were drafted to join our father in training sessions. A recently unearthed Polaroid shows us, impossibly tiny in diapers, standing alongside our impressively fit father in his swim trunks, all of us proudly grinning and arms akimbo in a superhero pose. It was 1979, the heyday of the movie *Superman*. All we needed were three capes to complete the look. *Am I fit or what?*

Every evening in the garage, the three of us moved in formation: forward kick, side kick, roundhouse kick. Our father would ask us to hold down his legs while he did sit-ups, or Andy and I would dangle from his biceps like a pair of baby monkeys while he lifted and swung us. After dinner, under the yellow sodium glare of the neighborhood streetlights, we'd flank him on nighttime jogs down to the parking lot behind our pediatrician's office, a mile away. We'd chase lightning bugs, and our dad. What did we learn, as children, from all of this early training? That nocturnal exercise was normal in our family, though not so normal in others. And that being strong was good, for each of us alike.

Exercise was fun in our house, because our father was a perpetual kid, wonderful at playing. Certainly, there was a measure of vanity involved. His was a febrile imagination; as he molded us into miniature versions of himself, he enjoyed the fantasy that he could live forever through us, his modest experiment in immortality. "Pick a sport," he said. First, we tried soccer, which didn't stick; then swimming, which did.

It must be said, though, that there came to be an unacknowledged specter over the whole muscular enterprise, and that was my father's father. He was the one who had instilled the value of early exercise in my own father, and yet he died suddenly, unexpectedly, from a heart attack at sixty-four. I was eight years old. I remember the phone call that delivered the news as a kind of shock wave. My mother was quiet, worried. And my father has been preoccupied with outrunning death ever since.

By the time I was in high school, my dad spent more of the year in Hong Kong than at home in New York with us. Slowly, and then all at once, he stopped making the return. The heavyweight anchor of my childhood disappeared. But I kept lifting and stretching and moving, in pursuit of the life of physicality he introduced to me—and, as part of the same inheritance, to ward off the specter of death, too.

AT THE MOST basic level, muscle powers and animates our existence.

The biologist and biomechanics pioneer Steven Vogel wrote that "muscle has been our sole engine for most of our time on earth." He pointed out that whether it's the tiniest flea or the largest whale, what moves and propels creatures great and small is, well, "the same stuff." Evidence of animals first flexing their muscles dates back 560 million years, to a recently discovered fossil of a cnidarian, an animal phylum that includes modern jellyfish, corals, and sea anemones. It has bundles of muscle fibers arranged in radial symmetry.

When we talk about what moves us as human beings—if you really want to get down to the heart of things, the meat of it, figuratively—it's muscle, literally. Strongest and biggest muscles? In your heart and your jaw, and in your butt. (We'll talk more about that later.) Smallest and weirdest? In your ear is the stapedius, just one millimeter long, controlling the vibrations of the stapes, a.k.a. the stirrup, the smallest bone in the body. And perhaps there are muscles you've never heard of, teeny ones in funny places—like the arrector pili, the little muscle fibers that give you goose bumps. Maybe you have them now, just picturing them.

Cardiac, smooth, skeletal: These three different types of muscle make our hearts beat; push food through our intestines, blood through our vessels, and babies out the uterus; and attach to our bones and help us get around. Skeletal muscles are the ones we move at will; the others work under our bodies' control, without our conscious effort. Individually, they do different things. Collectively, they drive us through our days.

Muscles deserve more consideration than we give them. We often think about muscle as existing separately from intellect—and maybe even oppositional to it, one taking resources from the other. The truth is that our brain and muscles are in constant conversation with each other, sending electrochemical signals back and forth; our long-term brain health depends on muscles—and moving them—especially when it comes to aging bodies. But the closeness of muscle and mind is not just biological.

Being a writer as well as a lifelong athlete, I can't help but notice how language is telling. *Muscle* means so much more than the physical thing itself. We're told we need different metaphorical muscles for everything: to study, to socialize, to compete, to be compassionate. And we've got to *exercise* those muscles—putting them to use, involving them in a regular practice—for them to work properly and dependably.

We flex our muscles to give a show of power and influence. We have muscle memory; it's a nod to the knowledge we hold in our bodies, of all things sensory, physical, and spatial. We lift ourselves up and jump for joy. We muscle through hard things;

that shows grit. Even when it's a stretch, we still try. And when we finally relax, it's settling in, acceptance, letting go.

The way you build muscle is by breaking yourself down. Muscle fibers sustain damage through strain and stress, then repair themselves by activating special stem cells that fuse to the fiber to increase size and mass. You get stronger by surviving each series of little breakdowns, allowing for regeneration, rejuvenation, regrowth.

Building up and breaking down: We exist in this constant cycle. In fact, it is the cycle that *allows* us to exist. When it ceases, so do we. As a species, we try to prolong the cycle, but we wrestle with the reality. The ancient Greeks saw the fit, well-exercised body as virtuous. But the flesh-and-blood humans that the immortal gods fell in love with were so handsome in form that they were sometimes punished for their beauty, however fleeting it always proved to be.

We move our bodies through the world, and our minds follow. The artist Paul Klee described visual art as a record of movement, from beginning to end—a drawing of a dancer, say, is made by a roving hand, which pins down the movement of said dancer, and the finished work is then appreciated by an audience's ever-tracking eye.

This book is an invitation to explore the many ways that muscle is the vivid engine of our lives. Note that this is not an anatomy textbook; nor is it a guide to working out. What you will find, though, are stories about the stuff that moves us and why it matters.

When I reflect on why I wanted to write a book about muscle, I realize that a lot of it has to do with a longing for my dad. I found myself wanting to write about things I can talk to him about. To go deeper into the muscle inquiry and pull him back into my orbit. To recover some sense of that closeness we once had.

Make me a muscle. A little girl sticking her arm out to make a muscle is funny. But over the years, the baby fat fell away and the muscles got stronger, and I found that instead of feeling funny, I felt fearless. Not completely, and not all the time, but what I came to understand was that what my father gave me was an understanding of my own potential. Everyone has been asked to make a muscle at some point, to demonstrate a whole host of things, tangible and intangible: strength, flexibility, endurance. Show me you're in good form; show me you're a person of action. Character that's grounded in something you can feel. It's a way to assert presence. To say: *We are here—conscious, corporeal, alive.*

This philosophy of muscle dates back to the ancients. Part of the fascination is the understanding that we are all eventually helpless before the cellular clock. Even as we alter our bodies and seek beauty and perfection. Even as we use muscle to acquire power and dominance that is political, economic, cultural, racial, or sexual in nature. Even as the effort to reshape verges into distortion and dysmorphia. Even as we stretch to our limits and seek transformation and transcendence. Even as we chase longevity, that forever-elusive immortality, we can only push so far before opposing forces yank us back.

But that doesn't stop us from trying.

STRENGTH

Muscles are in a most intimate and peculiar sense
the organs of the will.

—G. STANLEY HALL

1

What's Power, in a Body?

The original real-life Hulk was a mom with a baby.
 Or at least that's the story Jack Kirby, the creator of *The Incredible Hulk*, would come to tell. In a 1990 interview, Kirby described how he once saw a small child get stuck underneath the running board of a parked car. When the child's mother realized her son was in danger, her eyes widened in panic. She—an average mortal—raced around the car, grabbed hold of the rear bumper, and lifted the car off her baby.

"It suddenly came to me that in desperation we can all do that—we can knock down walls, we can go berserk, which we do," Kirby said.

"Whatever the Hulk was at the beginning I got from that incident," he continued. "A character to me can't be contrived. I don't like to contrive characters. They have to have an element of truth. This woman proved to me that the ordinary person in desperate circumstances can transcend himself and do things that he wouldn't ordinarily do. I've done it myself. I've bent steel."

The story of the Hulk, Kirby said, is really a parable about desperation. It gives a human being strength beyond their wildest imagination—the capacity to move heaven and earth, with ordinary muscles powered by extraordinary will.

Kirby wasn't exactly a feminist. Though he had witnessed a woman under duress demonstrate a controlled act of extreme strength, he cracked jokes about her body—"I'm not saying she was a slender woman"—and ended up making his Hulk a kind of Frankenstein, for whom intense emotion was a trigger to rage *out* of control. In that monstrous incarnation, the Hulk could never remember who he was.

For much of human history, female strength was deemed unnatural and uncanny, and the notion of it being tied to a crazed or highly emotional state of excitement is similarly ancient. Our species has a long history of thrilling true-life tales of what has come to be known as *hysterical strength*: extraordinary displays of human strength in response to dangerous or life-threatening situations. The *hysterical* part of the term traces back to the Greek *hystera*, which means "uterus." These days, hysterical strength can refer to episodes involving men, but what actually makes these stories most gripping, of course, is when they're unexpected. The two teenage girls who lift a tractor that has tipped over on their father. The mother who fights off a polar bear threatening her young son and his friend. The college student who throws a car off her unconscious father, who was crushed underneath when the jack collapsed. She pulls him out and performs CPR. He lives.

When Jan Todd was a girl growing up in Western Pennsylvania in the 1950s, she heard a version of what is now the clichéd tale of hysterical strength: An accident, a child trapped underneath a car. A mother, managing to escape the vehicle, lifting it up to save her child. At the time, these kinds of stories had the whiff of the apocryphal to her—something outside the realm of possibility, or, at best, a mostly true tale that was made bigger and taller for personal gain.

Little did she know that she herself would be capable of so much more.

Todd eventually went to college in Macon, Georgia. By the time she was twenty-five, she held three world records in powerlifting and *Sports Illustrated* had profiled her as the "World's Strongest Woman." Todd set her sights on Scotland, perhaps the epicenter of stone-lifting culture: The rocky landscapes of the Highlands are littered with remnants of ancient strength tests, with stone-lifting games held there every year as part of traditions that extend back as far as 2000 BCE. Tests of strength were done ritualistically in front of peers. In Gaelic culture, heavy lifting stones, or *clachan togail*, were sometimes used to prove manhood. These were often spherical granite boulders that were difficult to lift because of their size and smoothness, but they could also be of irregular and awkward shape.

Nowadays, the most well-known of Scotland's lifting stones are the Dinnie Stones, named after Donald Dinnie, who famously spurred the revival of ancient Scottish stone-lifting culture in the 1800s by performing the extraordinary feat of lifting and carrying two massive, rough-hewn granite boulders across a bridge. The Dinnie Stones have a combined weight of 733 pounds.

In 1979, Jan Todd became the first woman ever to lift the Dinnie Stones. For nearly four decades, she would remain the only woman to accomplish that feat. Every once in a while, at home, in the driveway, she'd pick up the side of her Ford Fiesta—sometimes for an audience, but mostly for herself.

I think of these feats of strength, hysterical or not, as stories of someone in transformation. They show how the spirit moves us to move.

IN BIG WAYS and small, life is a movement-based relationship with everything around us. Muscles make my fingers fly across these keys, knit my brow in concentration, correct my seated posture, shift my gaze to the window, square my shoulders, tap out the rest of this sentence. So much has become virtual, and yet my body still very physically influences my thoughts even as it conveys them to you. Your own muscles allow your eyes to take this in, to blink thoughtfully and tuck your chin in hand and tilt your head in consideration. We haven't said a word, but our bodies are talking to each other—even through the page (or screen, or audio recording, for that matter).

Mind over matter. Mind over body. What if I told you that the mind exists *only* to move the body?

The British neuroscientist and Columbia University professor Daniel Wolpert has a favorite story he tells about why we and other animals have brains: The humble sea squirt, a small marine animal, swims around the ocean as a tadpole-like larva until it finds a hard surface to alight upon. As soon as it makes its selection and attaches itself to that surface, the sea squirt—well, it liquefies and digests its own brain and nervous system for food, rearranging its organs and leaving just a bit of nerve tissue for the rest of its anchored adult life. It no longer needs the luxury of a brain or a nervous system because *it no longer needs to move.* Sea squirts produce both eggs and sperm that they release by spawning; some species can also reproduce asexually, by budding off clones. Brains are important only in that they allow us to move, interact, and exert our influence on the world, Wolpert explains to me. And muscles make that happen.

Exerting our influence on the world: That's the modern-day definition of a *flex*.

The evolutionary idea that we move to live is not so hard to understand. Even the earliest animals on Earth, with their rudimentary bundles of muscle fibers, flexed those bundles for a reason. We move to find favorable environments, abundant food, secure shelter. Consider the arctic tern, perhaps the ultimate endurance athlete, flying between the poles, twenty-five thousand miles a year. The tern's life is spent chasing endless summer from the Arctic to the Antarctic and back again, pinballing from continent to continent as good weather and food dictate, always returning to its home nesting grounds in the High Arctic.

Animals attract mates with potent displays of reproductive fitness that require considerable physical effort—a mesmerizing dance and an offering of food, perhaps, or the building of a beautiful nest. I think of the male white-spotted pufferfish, who spends a week or more bulldozing sand along the seafloor with his head and fins, all to create a magnificent flowerlike pattern twenty times his size that will draw the eye of a female. An approving female will swim to the center of the mating circle—an elaborate, whorled structure that the male puffer has adorned with seashells—and lay her eggs there.

Strength is one of the primary ways to demonstrate health and worth to a prospective mate, and to establish social hierarchy. Male fiddler crabs have one larger claw—a.k.a. the major claw—that they wave with power and speed, at great caloric cost, to show dominance. For a male silverback gorilla, chest drumming, strutting, and aggressive acts like throwing, hitting, and kicking are used to reinforce control in a group, or to recruit members to a new group. Birds of prey are one group of animals in which females are known to be larger and stronger than males; though the females don't initiate courtship in all raptor species, many do, and scientists theorize that their larger size—sometimes a third larger than their male counterparts—also helps them to catch and kill more substantial prey for food and to defend their nests and territory from threats.

What about humans? Up until about the age of ten, boys and girls have similar bodies and physical abilities; during puberty, a surge in testosterone signals the typical male body to grow rapidly, with significant gains in muscle and bone. On average, men have 80 percent more muscle mass in their upper body then women do, and 50 percent more mass in their legs. As with other primates, David Epstein explains in his book *The Sports Gene*, the larger size, longer limbs, bigger lungs, and greater blood volume of early ancestral human males evolutionarily translated to a higher capacity for the physical work of survival—the running, hunting, and fighting for mates that characterized human life at that time.

Even in modern times, without natural selection being the primary driver, the public showcase has historically skewed male when it comes to the human performance of strength. Across countless centuries and civilizations, our mythologies of manhood have been inextricably tied to it. What happens when a woman steps in?

IT WAS SPRINGTIME, 1973, and, in a sun-dappled Georgia meadow next to a stack of logs, a young philosophy student named Jan Suffolk was about to set foot on the path that would lead to her becoming the world's strongest woman.

The stack of logs and the meadow belonged to a good-natured bear of a man named Terry Todd, and the setting was an end-of-season party for faculty and students who played on the intramural softball team at Mercer University. Several guests sat atop the pile of logs, which had been culled during a winter storm. Over beers, the conversation turned to a traditional Scottish feat of strength called the caber toss.

Todd was a professor of education who founded the African American studies program at Mercer, but he was also a former junior national champion in Olympic weightlifting and a super-heavyweight national champion in powerlifting who wrote his doctoral dissertation on the history of resistance training. In other words, he knew the iron game. He enthusiastically explained the caber toss to his guests: Imagine taking a heavy, tapered telephone pole of a log—a.k.a. the caber, from the Gaelic word *cabar*, which means "rafter" or "beam"—and standing it on its smaller end. Now imagine picking it up, with that end cradled in your hands against your chest, and then running forward with the log; as it begins to fall, you hurl it up in the air, end over end, so that it lands on its heavier end and falls to the ground with the smaller end now pointing away from you, as if at noon on a clockface directly opposite. The straighter the throw, the greater the distance, and distance is what wins the event.

The conversation about the caber toss quickly became more than a conversation, and, as often happens at a party, several men stepped up to demonstrate something they'd never done before. Being the resident historian of physical culture, Todd helped to select the proper caber from the woodpile, and partygoers gathered to watch and hoot. As one of the men struggled in his repeated efforts to throw the caber, a young student with freckles and long blond hair emerged from the group.

With little fuss, Jan Suffolk stepped up, in her jeans and tennis shoes, stood the log on its tapered end, lifted it, and, with a few steps forward for momentum, deftly flipped the log over so that it landed pointing straight away from her at twelve o'clock. The guests erupted in cheers.

"As near as I can tell, that was the day I began to love her," Todd would later write.

Within a year, Jan and Terry Todd were married. The first thing Terry told his sister about meeting Jan would go down in family history: "You know, she's perfectly leveraged for the squat."

THE SQUAT IS one of the three classic exercises that make up the sport of powerlifting, the other two being the bench press and the deadlift. The primary muscles involved in the squat include those of the hips, the legs, and the butt: With a weighted barbell placed on the shoulders, you lower yourself down as if to sit on an imaginary chair, then stand back up. The bench press, on the other hand, engages the upper body muscles—prime movers in the chest, the shoulders, and the arms: You lower and raise a barbell from the chest while lying flat on your back on a bench. The deadlift is the most basic and elemental of all: Bend over, pick up a heavily loaded barbell from the floor, and stand up. It largely uses the muscles of the legs, the back, and the butt. As a set of three, the exercises represent functional strength and all-around physical power.

Hence the name *powerlifting*.

"Those for whom happiness is doing or watching the power-lifts, all I can say is that they—we—have chosen a sport which is as basic and ancient as any in the world," Terry wrote in the influential book *Inside Powerlifting*. As a sports academic and a former competitive lifter, Terry understood both the storied traditions behind powerlifting and the physical effort required to practice it. As a coach, he also had an internal radar for strength potential in others.

In photos from the 1970s, the Todds are radiant in their hale and hearty youth. At a bearded six foot two and 245 pounds, Terry's strength was apparent: legs the size of tree trunks and a barrel chest broad enough to hold a place setting for dinner. He'd begun lifting weights as a twiggy teenage tennis phenom, with the goal of strengthening his nondominant, left arm. Though most sports coaching in the 1950s and '60s was decidedly against weight training—it was thought to limit range of movement, speed, and agility, rendering an athlete "muscle-bound"—Terry came to realize that he could jump higher when he weighed 340 pounds than when he weighed more than a third less.

It seems obvious today, when strength and conditioning coaches are de rigueur for any sports team, but the idea that weight training could contribute meaningfully—and functionally—to athletic performance would remain outside the mainstream until the mid-1970s. The squat, the deadlift, and the bench press are today considered three essential resistance exercises for training athletes of all kinds. The Todds would play a major role in pioneering that practice—and in documenting the change.

"WHEN TERRY AND I met, I didn't know anything about that world," Jan tells me.

I've come to visit her at the University of Texas at Austin, where she is chair of the Department of Kinesiology and Health

Education and the director of the department's PhD program in the sport humanities. It's also where Jan and Terry founded the H. L. Lutcher Stark Center for Physical Culture and Sports—a museum, a library, and an academic research collection on the fifth floor of the university's hallowed football stadium, above the Athletics Hall of Fame.

Terry died in 2018, but his influence is felt in every detail of the center. It includes rare photographs, papers, artifacts, and more than forty thousand books on the subject of physical culture, with a focus on the world of strength: sports training, weightlifting and powerlifting, strongman contests, historical feats, the Olympics, and more. It is recognized by the International Olympic Committee as an official IOC research center, one of only three in the United States.

The Stark Center is also the official archive of UT Austin's intercollegiate athletics program, documenting the university's greatest coaches and student athletes. Between 1986 and 1996, the Todds themselves coached the Texas Longhorns to nine men's and women's national powerlifting championships, plus four overall combined team trophies.

A record-breaking strength athlete who went on to become one of the foremost academic authorities of physical culture and sports? I admit that the appeal of the superhero double identity called me to seek Jan out. Here was someone who'd revealed herself to be unusually, thrillingly strong when no one had expected—or particularly wanted—her to be. Someone who could talk about what it felt like to occupy such a body, and also what it meant from a societal standpoint.

What makes one person stronger than another? Scientists have identified a number of genes that are involved in muscular development and growth, but exactly how these affect an individual's muscle strength is unclear. For example, the MSTN gene encodes for myostatin, a protein that inhibits skeletal muscle growth; lower levels of myostatin mean *more* muscle mass. Natural variation in myostatin allows some people to beef up more easily than others, but it's not the only thing that determines strength. The same goes for genes regulating testosterone; men generally have more of this hormone than women, but not always. Just as some women are taller than some men due to natural variation, some women have higher levels of testosterone than some men.

Recent research on testosterone-induced muscle growth also reveals that not all muscles respond the same way to hormones. And while testosterone gets more attention, the hormone estrogen also dramatically affects muscle function for both men and women—boosting growth and strength, helping to repair fibers and reduce injury, and regulating metabolism.

What we do know is that a complex dance of genes, biology, and environment shapes our physical abilities. And strength, of course, is so much more than physical—it's psychological too.

Jan herself is unaware of any biological quirks of her own, and she is the first person to express surprise at how her life has revolved around muscle to the extent that it has. Back in college, she was tall and agile, but she didn't compete in organized sports—it was the pre–Title IX era, when opportunities for girls to play sports were few and far between. She was the fastest girl in her high school, but the school didn't have a track team. "My grandmother would tell me not to run too fast," Jan explains. "She'd say, 'You don't want the boys to think you're faster than they are.' My father wouldn't even let me do ballet, because he thought it would make my legs too muscular."

There's a story Jan tells about her father that reveals the shape of the world in which she was raised. A couple of years before she met Terry, her father visited during her freshman year at Mercer; that Christmas, he sent her a plane ticket to visit him in Chicago. She hadn't seen him much since her parents' divorce several years before. At the Field Museum, they entered an exhibit filled with old-time carnival strength machines.

Jan's father was a powerful forty-five-year-old former Pennsylvania steelworker who weighed more than two hundred pounds. Her first memory of muscle, in fact, is of her dad showing her and her sister his biceps and playfully making it jump. He examined one machine, a hand dynamometer, and gave it a squeeze. It registered a respectable grip-strength score—safely away from the pipsqueak end of things, closer to he-man status.

Then Jan tried it.

"I'm not sure who was more surprised, since at the time I was only eighteen and had never touched a barbell," she would later write, about besting her father that day.

In retrospect, it was the first inkling she had of her own nascent strength.

"So we tried it again," she wrote, "and got essentially the same results. I remember him speculating that the machine must be broken, but we didn't talk about it much after that as it was clear that he felt somehow undone by the situation."

Undone by the situation.

In the years since, Jan has thought a lot about that moment. "When he didn't succeed at that machine, it sobered him," she says. The visit was fraught in many ways, but when she looks back, she sees the beginning of her father's decline, and the long tail of his absence from her life; they became estranged soon after that.

But I see something else, and that is Jan's own ascendance.

"Muscle is muscle," she tells me. "What's different is the permission that society gives us to use it."

2

Muscle as Potential

As a show of strength, perhaps there's nothing simpler than picking up a heavy object. In ancient Egypt, hoisting a sack of sand and holding it overhead was a common practice; murals depicting the exercise can be found in the tombs of Beni Hasan, a vast necropolis carved into a steep limestone hillside on the eastern bank of the Nile River, dating back several thousand years.

Sometimes a heavy object held more rarefied meaning. The Han dynasty historian Sima Qian, who wrote the foundational history of China covering the two and a half millennia leading up to the early first century BCE, described the lifting of a three- or four-legged cauldron known as a *ding*, which could weigh several hundred pounds. Dings were once largely ceramic vessels used to prepare food, but beginning in the Shang dynasty, around 1600 BCE, they were cast in bronze and buried with their owners. They had ritual significance and were the preeminent symbol of wealth, divinity, and power; during the Han dynasty, ding lifting was included in the One Hundred Games, a display of Chinese martial arts, acrobatics, music, and dance, performed only for royal and elite audiences.

When I ask my father, who lives in Guangzhou, about this history, he tells me that Sima Qian's most famous lifting story involves death by ding: King Wu of Qin, a great admirer of strength who elevated strongmen to his royal court, broke his leg while attempting to lift and carry a particularly consequential ding. He later died of his injuries, throwing the Qin state into disarray.

On a visit to the Metropolitan Museum of Art in New York, I speed-walk directly to the China galleries to examine several of these squat metal cauldrons, some of them inlaid with pigment or oxidized green. The bigger and more elaborate the ding, the richer and more revered the family.

In many cultures, of course, specialized objects were not necessary for lifting—any heavy rock would do. Even the Old Testament points to the lifting of stones as a customary trial of strength. According to Saint Jerome, the fourth-century priest and biblical scholar known for his Latin translation of the Bible from Hebrew and for his accompanying commentary, the description of Jerusalem as a "burdensome stone" is a nod to the ancient practice of stone lifting, which was still common in his time, in cities and villages across Palestine and Judea.

In these places, Saint Jerome wrote, "round stones of very great weight are placed, at which the youth are wont to exercise themselves, and according to their differing strength to lift them, some to the knees, others to the navel, others to the shoulders and head; some exhibiting the greatness of their strength, raise the weight above their head with both their hands straight up." Saint Jerome also made note of the not-insignificant risk involved of the stone falling and "crushing them to pieces."

In Japan, thousands of *chikaraishi*, or strength stones, can be found across the islands, dating back to at least the eighth century;

many chikaraishi are located at Shinto shrines and temples, and were likely used in divination ceremonies. Stone lifting was the rare traditional sport practiced by the samurai and peasant classes alike.

In Icelandic culture, lifting stones were once used to qualify men for work on fishing vessels; a heavier lift would ensure a man a greater share of the catch.

Muscles also drive the narrative in a Hawaiian legend involving King Kamehameha I, who united the islands into a single kingdom in 1810: When Kamehameha was a teenager, his strength was revealed and a prophecy fulfilled when he lifted and flipped the Naha Stone, a slab of volcanic rock that reportedly weighed seven thousand pounds. Today, you can find the Naha Stone on the Big Island of Hawaii, right in front of the Hilo Public Library. (No word on whether anyone has tried to lift it recently.)

The bottom line: Lifting a heavy thing is atavistic. It has long conferred status, resources, land, women, rights. Lifting an *important* heavy thing showed a man to be a warrior, capable of defending his people and property. Perhaps you could say it was a primitive way to identify what we now call leadership skills. But the traditions of stone lifting persist to this day, from Iceland and India to Scotland and Spain. Feats of strength still capture our imaginations: *Look, and be amazed!* Strength is something to be celebrated.

The more I think about it, the more I find that the idea of strength as a proxy for worthiness, ability, or success has interesting legs. A friend told me about a venture capitalist she met at a cocktail party. When she asked him how he decided which people to give money to—in the uncertain land of start-up businesses, where everything is a gamble—he told her that he liked to invest in athletes. It was a kind of shorthand, he said, for finding people with fortitude. "Athletes understand how to push themselves past the point of pain," he explained, when other—presumably lesser—beings might give up.

Even today, we view physical strength as a positive representation of character. In a world where we are living increasingly virtual lives, I'm surprised by the idea that power of all kinds—political, economic, and social—can still originate from the physical. Just look at former governor of California Arnold Schwarzenegger, who first came to fame as a bodybuilder known as the Austrian Oak, with muscles that "jump out at you like outrageous price tags." (Schwarzenegger's chief of staff tells me that his boss still goes to the Gold's Gym in Venice "all the time," to stay connected with "the guys"—his original and most loyal audience.) Or Dwayne "The Rock" Johnson, a former college football star and pro wrestler, who is now one of the highest-paid actors and entertainers in the world; he, too, has expressed interest in politics, with a potential future run at the presidency. (The Rock's personal gym goes wherever he goes; nicknamed the Iron Paradise, the gym includes twenty tons of equipment that must travel by dedicated eighteen-wheeler.)

Choosing people based on some shared trait, like strength and athleticism, is tribal. But there's something specific about strength itself that I want to unpack. Physical strength has been good for men; for women, less so. What strength means can vary by individual; when we say someone is *too* strong or *too* muscular, it's often a comment on what we permit that person to be in society.

Maybe that's what makes some people uneasy: muscle as potential. And sometimes we don't know our own power, until, finally, we are given the opportunity to discover it.

JAN TODD MAY not have known her own strength, but she was on her way to figuring it out. On one of their first dates, Terry brought two bottles of beer out to the backyard. He idly put a bottle cap between his thumb and first knuckle and bent it in half. "You want to try?" he asked Jan.

She did it easily, so he upped the ante. He repeated the trick, this time with thumb and index finger straight. The decrease in leverage meant that the task required significantly more grip strength.

"Sure, I'll give it a try," Jan replied—and promptly pinched the cap closed.

Terry knew that it was a rare person who could perform this feat without training. After all, he'd studied the history of strength training, everything from grip-strength tests to early twentieth-century strongmen and strongwomen. This was sword-in-the-stone territory, the stuff of Arthurian legend. What else was Jan capable of?

He kept his thoughts to himself. "When we married, there was no plan for me to be a weightlifter," Jan tells me. "In those early days, we were just doing hippie stuff—we lived on a small farm, we had goats and rabbits. I was busy with my studies." Though he'd retired from competition, Terry still lifted weights every once in a while, and Jan began to accompany him to the gym, using light dumbbells to improve what she'd always felt was a round-shouldered posture. This was an era when weightlifting wasn't the norm, much less women in the gym.

During one Christmas visit to Terry's family in Austin, the two were working out at the Texas Athletic Club when a young woman walked into the gym and began a series of deadlifts.

Remember that in a deadlift, a loaded barbell is lifted from the floor to mid-thigh, and the movement engages a cascade of muscles in the legs, the back, and the butt. The quadriceps—the muscles of power at the front of the thigh, used to straighten the knee—figure prominently. The deadlift is a simple measure of brute strength, what sportswriters have called "the great separator." It doesn't require special technique—unlike the clean and jerk in Olympic weightlifting, which involves carefully calibrated multistage movements—so it is power distilled.

The woman Jan observed at the club was small in stature, but she kept adding weight to the bar until she maxed out with a lift of 225 pounds—twice her own body weight. Jan was mesmerized. She'd never tried the deadlift, and she wandered over to chat with the woman. By the end of the afternoon, Jan herself was deadlifting 225 pounds.

"When I lifted that first time—when I was really trying to pick up something heavy—it was a different feeling, a powerful feeling," Jan says. Being strong, she was beginning to find, was immensely satisfying. "And on the way home, Terry told me, 'You know, there used to be women who did this. You could do this, too, if you wanted to.'"

Terry shared the story of Katie Sandwina, a professional strongwoman in the Ringling Brothers Circus, who, in the early 1900s, became famous for carrying a 600-pound cannon on her shoulder and juggling her 160-pound husband, Max, as part of her act. Sandwina would do this while wearing tights and a corset, her hair in ringlets.

This was during the golden era of professional strength athletes, when strongmen and strongwomen emerged as popular acts on the circus and vaudeville circuits across Europe and the United States. Dramatic color posters from the period depict Sandwina as a majestic Valkyrie figure on a horse-drawn chariot—carrying a spear and wearing a winged helmet, leading men into battle—and in leopard-print or mini-skirted outfits that showed off her muscular arms and legs as she held anvils on her chest or hoisted multiple men in the air. (Examples of these stunning posters can be seen hanging in the Stark Center galleries today.)

Terry had accumulated an impressive personal collection of old books and magazines at his parents' home, and, somewhere in the middle of an impromptu crash course in the history of female strength, Terry and Jan came across a copy of *The Guinness Book of World Records*. One entry in particular caught Jan's attention: "The highest competitive two-handed lift by a woman is 392 lbs. by Mlle. Jane de Vesley (France) in Paris on October 14, 1926."

For Jan, the idea that she could be strong—and not just strong, but exceptionally, *record-breaking* strong, part of this admirable lineage—shifted her perception of what was possible for herself. She turned to Terry and smiled. "I think I can beat that," she told him. They agreed that it would be a fun goal, because wouldn't it be a hoot if she made it into *The Guinness Book*?

The two went back to Georgia and got to work: twice-a-week weight workouts that focused on the squat and the deadlift as the two primary exercises, plus the bench press to strengthen Jan's upper body and grip. On other days, she'd run a mile and throw in a few wind sprints for conditioning.

Sixteen months later, on May 3, 1975, at a meet in Chattanooga, Tennessee, with a lift that sent the crowd leaping to its feet with excitement, Jan set a new world record for the two-handed lift by a woman: 394.5 pounds.

Over the next decade, Jan would set world records in five different weight classes, ranging from 148 pounds to 198-plus pounds; win the first women's powerlifting championship; and become the first woman to total more than a thousand pounds in the three powerlifts combined. In 1977, *Sports Illustrated* declared Jan to be the strongest woman in the world, with the following (wonderfully specific) criteria: "if the strength being considered is muscle strength and if it is measured in units of heavy iron."

The article also did something that, to me, seems groundbreaking. It featured a woman talking about how good it felt to be strong, and how much pleasure she took in that power. "I love the way it makes me feel," Jan told the magazine. Lifting, she explained, had expanded her worldview.

Soon after, Jan appeared on *The Tonight Show Starring Johnny Carson*, instructing Carson on how to lift a barbell loaded with 415 pounds, which, as the footage shows, he is laughingly unable to budge even an inch. His facial expressions and body position telegraph delightful incredulity. He finally lifts the bar in a group effort with the help of two other guests, the actors Carl Reiner and Jack Klugman. Jan, of course, does it solo, performing three reps with a weight that's more than the official world record at the time.

In the wake of all these achievements, and with newfound fame outside the relatively insular world of powerlifting itself, Jan was often asked to demonstrate feats of strength, much like the strength athletes of Katie Sandwina's day. The *Guinness Book* people invited her to perform at state fairs; if she was appearing at an exhibition, say, she might lift some barbells. But she and Terry had fun coming up with new ideas, often inspired by the strongmen of yore. She would bend metal spikes, drive a nail through a board with her bare hand, or pick a dozen children from the audience, put them all on a big table, and lift the table from the floor on her back. She called this "the kid lift."

Metal spikes, nails, the kid lift: This was a strength language that people got instinctively, and they loved it. In the telling of these stories, Jan shares that she and Terry had the time of their lives exercising their creativity. With these objects, the audience could comprehend in a real way how heavy things were—not from the number stamped on an iron weight, but from the object's relationship to the average person.

A big, unwieldy rock, or maybe two big rocks at once? That was something everyone could understand.

3

A Heavy Lift

Two famed granite boulders lured two of the most accomplished powerlifters in history to Scotland: It reads like a line from a travel brochure. As Terry would eventually write in an article for *Sports Illustrated*, "In all my reading nothing seemed quite so wonderful as the tales of brawny Scots hauling huge stones from the heather." He'd long wanted to attend and compete in the Highland Games himself, but in the year of his peak fitness, the contest conflicted with his defense of the American powerlifting title, and so he missed his chance.

To Terry, the rich, colorful history and the longevity of the competition were irresistible—the Highland Games featured stone lifting, shot put, caber toss, hammer throw, tug-of-war, and footraces, all with dancing, bagpiping, and a convivial country-fair atmosphere. These days, Highland Games gatherings and events are held all over Scotland throughout the summer months. Though the games are indisputably of the specific geography in which they were born, their fame and influence have long been global.

At the 1889 World's Fair in Paris, Baron Pierre de Coubertin—the father of the modern Olympics and the founder of the International Olympic Committee—was in attendance. In the midst of his plans to revive the Olympics, the baron was so impressed by what he saw at a display of the Highland Games that he ended up introducing the shot put as an event at the first modern Olympics, in 1896; the hammer throw and the tug-of-war would soon follow. (Regrettably, the tug-of-war was only in official play between 1900 and 1920, but the shot put and hammer throw remain on the track-and-field program today. Note also that women were first allowed to compete in the 1900 Olympics, but there were only twenty-two of them, and their participation was limited to tennis, golf, sailing, croquet, and equestrian events.)

Jan had heard Terry talk often with Scottish historian David Webster, the foremost authority on the Highland Games. Webster had written a book about the games, which included a section on the revered Dinnie Stones, as well as a passage that seemed to be a direct invitation to Jan: "Maybe one day as athletic standards go higher still, we will have women attempting to lift the stones of strength!"

"I loved the idea of the Dinnie Stones," Jan tells me.

They were a difficult and unusual challenge, novel to her but also firmly tied to history. The two granite stones together weigh 733 pounds—the smaller stone 318.5 pounds, the larger 414.5 pounds—and a large iron ring is embedded in each. Local lore has it that Donald Dinnie was a young man when he first went along to help his father, Robert, a stonemason, make repairs to the Potarch Bridge, a stone arch that stretches over the River Dee in Aberdeenshire. The stones were used as counterweights during repair work, but mostly they were used to tether horses at the nearby inn. Sometime in the 1860s, Robert was the first to stand astride the stones and lift both of them from the ground at the same time.

As with any good yarn, the details tend to vary by who's doing the telling and how many pints have been shared at the pub. Suffice it to say that Donald Dinnie picked up the stones and reportedly carried them across the width of the bridge. In the following years, his fame would grow with legendary performances in the Highland Games themselves.

For her part, Jan knew that she could pick up the Dinnie Stones one at a time. But their large size, irregular shape, and the fact that she would have to straddle the stones to pick them up at the same time portended great difficulty. By the late 1970s, only a handful of men out of thousands attempting the feat had been recorded as successfully lifting the stones, among them Robert and Donald Dinnie; Jack Shanks, a Belfast police officer, more than a century later, who was the first to lift them without gear since the Dinnies did; a local man, Jim Splaine, who held the record for lifting them more times than anyone else (as Jan puts it, "He'd go to the pub, have a beer, pick them up, and go home"); and—surprise!—David Prowse, the actor who played Darth Vader in the original *Star Wars* movie franchise.

"He was six foot six, with legs for miles," Jan says. "That was not me—even on a good day, I was a foot shorter." She knew she needed to strengthen and add muscle to her back. "Whatever accommodations I had to make with my body in preparation"—e.g., gaining weight—"I was okay with that."

In the summer of 1979, she began to train at Auburn University, where she and Terry were helping to direct the newly established National Strength Research Center, where exercise scientists would study strength athletes in training.

She tried weird exercises, like the Jefferson lift, which required straddling a bar eighteen inches off the ground, gripping it with one hand in front of her body and one hand behind, and pulling it up until her legs were straight and locked. Her training log notes that she started with 500 pounds and kept adding weight until she could pull 805 pounds—more than the weight of the Dinnie Stones.

In the partial deadlift, which involves lifting a loaded barbell from knee level to thigh level with the muscles of the hips, thighs, and back, she began training with 600 pounds. By the time she and Terry left for Scotland, in late summer 1979, she was able to pull a whopping 1,100 pounds in the partial deadlift—the weight of a concert grand piano.

Later, Terry would describe their trip with plainspoken humor: Three Americans go to Scotland and lift some heavy rocks. The Todds' friend Bill Kazmaier, who at the time lived and trained with them and went on to become a world champion powerlifter, strongman, and professional wrestler in his own right, had joined the fun. He wanted to try his hand at various events at Braemar, the most famous of the Highland Games and the favored gathering of Queen Elizabeth and Prince Philip.

On a cool, misty afternoon under darkening skies, the trio headed to the Inn at Potarch for Jan to make her assault on the stones. It wasn't until they arrived there that Jan fully understood the challenge. The stones were significantly bigger than they looked in photos—"This was the pre-internet era," she says, "when you couldn't just look up pictures of five hundred different people with the stones for reference"—and the size of the larger stone would force her to take a much wider stance than was ideal for maximum leverage.

Terry was worried too. "The two brutal-looking rocks were chained together, and as I looked at them closely I began to fear, for the first time since Jan did 900 on the partial deadlift, that she might fail to lift them," he wrote in the article for *Sports Illustrated*.

Jan warmed up with the smaller stone, lifting it twice with each hand. In order for her to get better purchase, Terry and Bill moved the two boulders closer together, so that they were touching, in the hope that they wouldn't swing against her legs as she made the attempt. As a crowd began to gather outside the inn, Jan straddled the stones, grabbed the two iron rings, and pulled.

The small stone came up, but the big stone stayed put. She took a break, walking away from the spectators to regain her composure. The iron rings were slender, more so than a barbell, and lifting such massive weight attached to them cut into her hands. She used leather wrist straps to help with her grip, but the rings still hurt. The rocks, she thought, were just *so damn heavy.*

After a second attempt, in which the bigger stone still failed to budge, she took a walk with Terry over to the River Dee, staring for a moment at the flowers and the stone arch of the Potarch Bridge, where it all began.

Terry asked Jan if she wanted to give up the attempt—no lift, he said, was worth serious injury. But she decided that she wanted to try one more time.

He coached her as usual, reminding her about good foot positioning and leaning back to make room for the bigger stone to clear during the lift. But then he remembered something.

The previous evening, they'd gone to see David Webster, the Scottish historian, and they'd visited the home of a whisky distributor who knew of the party's mission to lift Scotland's most famous "manhood stones." The whisky distributor, clad in a kilt, had made a traditional presentation of gifts—a bottle of Scotch, a small scabbard knife known as a *sgian dubh*—to all the male guests in the party, to the exclusion of Jan. Given the primary mission of the pilgrimage, it was a strange slight, and one that had stung.

As Jan stood over the stones and prepared for her final lift, Terry whispered in her ear: "Let's see you pull this one for the whisky man."

Her face reddened as she yanked the smaller stone, which came up fast and high. The crowd yelled its approval as she leaned back farther—and the larger stone came up. A photographer snapped a picture of Jan, her face in profile, grimacing with the effort, long blond hair down her back, the two massive boulders clearing the ground at last.

As I write this, I consider again the list of advantages that lifting heavy things has traditionally afforded men. In assessing what lifting heavy things can confer to women, it occurs to me that there is one aspect that is perhaps most valuable of all: respect.

If Jan was hanging out with Terry and Bill, people would see two big guys and automatically assume that *they* were the champions. Then they'd look at her dismissively: *Well, who's the girl?*

"When I think back on all the stuff I did in those early times," Jan says, "in many ways I'm proudest of lifting the stones."

What strength meant to strongwomen from Katie Sandwina to Jan Todd was often much more than the diversionary spectacle that ordinary onlookers enjoyed. Lifting weights meant achievement by material increments. Not only that, but it was possible to build a life around it. For Jan, lifting the Dinnie Stones was a challenge that she didn't know she could meet. To show herself, and the world, what was previously unimaginable.

Jan pauses, thinking about Terry. "Having a historian in your life who remembered things," she says. "That meant something."

ONCE, JAN AND I swapped notes on our dads. Her father's rough dismissal of her participation in sports as a girl shaped the strong-woman she'd become, she said, "as much as if we'd played catch or if he had taught me to skate." The limits he placed, it turned out, were ones she would eventually be driven to breach. It wasn't as she might have wanted it, but it was true.

Conversely, perhaps the most striking thing about the physical education my brother and I received under the tutelage of our own father was that he trained us equally, without regard for size, age, or gender. He set us upon each other for sparring practice. Andy had entered the world one year, one week, and one day before me. If one of us kicked or punched the other to tears, my father would exclaim, "You forgot to block!" Then he'd laugh his uproarious laugh, dispense fierce hugs, and have us go another round. His approach had its flaws, but I grew up feeling that there was value in physicality, and that in this arena, I was limitless.

During my freshman year in college, I rowed crew. I worked hard to pack on the pounds to get myself to look less like a coxswain and more like a lightweight rower. It was the most time I'd ever spent in the weight room, and it has probably remained so since. I squatted, did bench presses, practiced lat pull-downs; I ran for miles along the Charles River in the snow, and lunged up and down the concrete steps of the Harvard University football stadium, where a full stadium tour is a leg-extinguishing thirty-five hundred steps. (Cue *Rocky* theme music.) When I went home on holiday break and picked up some hours lifeguarding at the pool, my friends whistled at the breadth of my back and shoulders. I'd be lying if I said that all of this wasn't satisfying on some basic level. I know from all three layers of my abdominal muscles that there's nothing like a daily dose of iron.

Weightlifting has also been shown to build resilience in the mind. New research with populations recovering from post-traumatic stress illustrates how lifting weights helps people feel more at ease and in control of their bodies. If you can improve your physical strength in a solid, visible way, it can reframe your self-perception: *Look what I did! This is evidence that I'm different now.* You can see yourself as a person with agency.

We always had a pull-up bar in our house. For a long time, it lived in the doorway of my father's downstairs studio. Every time he walked through, he'd do a set of pull-ups or a series of flips, like a gymnast on a high bar, before continuing on. For years, Andy and I struggled in imitation. We'd dangle from toothpick arms and thrash our feet, in the effort to get our faces ever closer to the bar.

Part of the fun of attempting the impossible, of course, is surprising yourself when you get there.

4

The Making of a Hero, Then and Now

On a blazing autumn afternoon in Texas, I head over to the Stark Center at UT Austin. On my walk across campus, I notice all manner of athletes: broad-shouldered swimmers leaving the pool, hair damp; runners at the track practicing starts off the block, rabbiting around the turns; towering basketball players loping across a grassy field. It feels serendipitous to glimpse these young bodies-in-training, flesh and blood working hard.

At the Stark Center, there are more fit bodies on display, their muscular forms arrested in sculpture, photography, and art, the first example of which greets me in the lobby: a massive ten-and-a-half-foot plaster cast of the *Farnese Hercules*, an ancient Roman statue that resides in an archaeological museum in Naples, Italy, based on an earlier Greek model. Hercules is majestic but weary, his phenomenally chiseled physique at rest after mythmaking labors.

This is a place where physical excellence is studied with intellectual rigor, and is elevated as such. Permanent exhibits feature such luminaries as the heavyweight boxing champion Joe Louis, nicknamed the Brown Bomber, whose 1938 defeat of Max Schmeling was celebrated as a triumph over Nazi Germany and racial bigotry; two weeks before the match, President Franklin D. Roosevelt invited Louis to the White House and told him, "Joe, we need muscles like yours to beat Germany." There's a display about the famed Muscle Beach strongwoman Abbye "Pudgy" Stockton, whose midcentury magazine column, Barbelles, encouraged housewives to get acquainted with their muscles—and demonstrated, with style and flair, that it was possible to be strong and feminine at the same time. And there's a gallery spotlighting the two-time Olympic gold medalist Tommy Kono, considered one of the greatest weightlifters of all time. Kono, who set world records in four different weight classes, began his athletic journey in the World War II Japanese American internment camp at Tule Lake, California; it was there, as a sickly child with asthma, that his neighbors introduced him to weightlifting. By the time his family was allowed to go home, three years later, he was fifteen, and he had gained about that many pounds in muscle.

As I wander around the galleries, I think more about muscle in terms of possibility: the meaning that muscle holds, and what it can allow us to accomplish in society, beyond the body.

Jan Todd has published dozens of scholarly articles, including papers on sports and exercise history, anabolic steroids, the science of strength, and the history of women and physical culture in the United States. In 1990, the Todds founded an academic journal, *Iron Game History*, which is published twice a year. Jan's numerous projects in progress include a book on the history of professional strongmen and strongwomen and a historical paper examining the little-known cultural currents informing Jean-Jacques Rousseau's eighteenth-century conception of men's physical education and the ideal body. "I'm interested in all the backstories," Jan says. "Sometimes it takes decades, but I always find myself going back and digging them out."

In the Stark Center archives, collections include every issue of *Sports Illustrated*, the personal papers of fitness guru Jack LaLanne, and a copy of *De Arte Gymnastica*, a sixteenth-century book that is described as the oldest treatise on sports medicine. In an ornately framed painting behind the reference desk, a young Arnold Schwarzenegger holds a classic Mr. Olympia bodybuilding pose (who can forget those muscles jumping out at you like outrageous price tags?).

I admire a hand-stitched leather medicine ball. Then I spot a massive dumbbell, partially filled with lead, that was custom-made for the early-twentieth-century Coney Island strongman Warren Lincoln Travis. At 1,560 pounds—that's the weight of a juvenile elephant!—the giant dumbbell is the single heaviest object on display at the Stark Center.

With all of these rare holdings, rules are required to protect against human impulse. "Our galleries include a number of weight-lifting equipment items—please do not touch or attempt to lift any of them," a notice says. Despite good signage, it appears that many of us cannot resist the attempt. Jan had invited me to the Austin Film Festival premiere of a documentary about Terry, and to attend the weekend's international Rogue Strongman Invitational, for which she is the event director. Many of the people strolling through the galleries this evening are *actual* strongmen—and, given the chance, they'd probably give a pretty good go at lifting all the heavy objects.

What is it like to hang out in a galaxy of giants? For me, a mortal of precisely average size and strength, it is to be in a constant state of awe. (Not to mention a constant state of looking up, up, *up*.) To spend a week in Austin mingling with strongmen old and new is to understand Jan's place in the firmament. At sixty-nine, she moves easily in her orbit; this is an ease she has earned, and everyone knows it. She greets friend and stranger alike with a warm embrace—"I'm a hugger," she says, with a note of apology. (Her grip: solid.)

At one point, Jan introduces me to the Icelandic strongman, actor, and boxer Hafþór Björnsson, famous the world over for his role as the Mountain from *Game of Thrones*. From his six-foot-nine perch, the Mountain smiles a shy half smile down at me as he shakes my hand, or, more accurately, engulfs it in his; with his other hand, he cradles the entirety of his year-old son, Stormur.

Later, Jan tells me that the Mountain has Terry's motto tattooed on his twenty-three-inch calf: DON'T WEAKEN. The full expression being *It's a good life if you don't weaken*. It was something his grandmother used to say.

I write Jan as a hero. She belongs in the company of such legendary figures as the French wrestler and actor André René Roussimoff, better known as André the Giant, and the Bulgarian-Turkish Olympic champion weightlifter Naim Süleymanoğlu, who, at a diminutive four foot ten, was pound for pound the strongest man in the world, and was celebrated as the Pocket Hercules—a nickname Terry coined for him. Terry's coverage of sports connected latter-day strength athletes—man *and* woman, big *and* small—to a once-forgotten history.

I write Jan as a hero because the stories we tell matter. The journalist Kate Fagan, who has spent her career studying the place of women's sports in society, has suggested that "the history of men's sports is uninterrupted mythmaking, the kind through which momentum is created." She points out that the transmission of these stories perpetuates ideas of greatness. "One generation to the next," she wrote in an essay for the *New York Times*, "we've heard the stories of Babe Ruth and Jim Thorpe, of Shoeless Joe Jackson and Jesse Owens. These stories are told, and retold, passed down through movies and documentaries and photography."

The lack of mythmaking for women athletes until recently has contributed to the fiction that women athletes are not worthy of the same attention. "Of the many explanations that exist for why men's sports are more popular than women's, the most prevalent is that men run faster and jump higher," Fagan wrote. "Ergo, men are more exciting to watch. This isn't a meritless argument; it's just simplistic and incongruent with—and this is just one of many examples—our obsession with the Little League World Series.

"What we don't consider often enough is how stealthily men's sports intertwines with history."

Little by little, this is changing. The urge to do something mighty is ancient. There's a nobility to it. To make your own attempt connects you to that past. As a weightlifter, sports historian, and director of the Stark Center, Jan shoulders and carries on the work of mythmaking. Which, if you think about it, is its own Herculean labor—of body and mind, together.

AS WITH SO many other athletes these days, the strongman's body has been professionalized, the training and the lifting and the eating regimented for maximum success. But even the eating has an element of myth to it. The *Irish Times* sportswriter Malachy Clerkin told me a story about Pa O'Dwyer, Ireland's strongest man five times over.

As Clerkin memorably put it, "The eating is a torture, a seam of misery running through O'Dwyer's day from long before dawn until well after dusk."

In the weeks leading up to a competition, a strongman like O'Dwyer needs to hit ten thousand calories a day—four times that of a typical male in his thirties—with the sole purpose of feeding his abundant muscle mass and maintaining his three hundred-plus pounds. Sometimes he would eat eight steaks a day, the last at midnight. With the size of his muscles—which require extraordinary amounts of energy to maintain, even while asleep—O'Dwyer could lose up to a pound overnight.

"So that's why I have to keep eating that late at night," he explained to Clerkin, "even though it's a pain in the hole."

At that point, he said, he hadn't been hungry in seven years. (Perhaps this is a good time to tell you that the masseter, a jaw muscle used for chewing, is the strongest by weight, able to exert up to two hundred pounds of force on the molars.)

For all its nobility, the pursuit of mightiness remains grounded in the body and all of its appetites. But the strength community's insatiable curiosity about the human body is something I find surprisingly moving. *To know one's own strength:* I've come to understand the meaning of these words not as a binary statement, an "I do" or an "I don't," but as an ongoing process of discovery. Muscles matter—they allow us, in an observable way, to see what we *can* do. Though you may not initially know what you're capable of, you have vast reservoirs of potential, waiting to be tapped. For just the right moment to be revealed.

JAN GOES BACK to Scotland most every year, to attend the Highland Games in Aboyne, around which the Dinnie Stones are featured in strength events. There is a new women's event named after her: the Jan Todd Classic. Even now, she speaks with a trace of disbelief about what she has accomplished as a champion powerlifter and a groundbreaking scholar of physical culture. "It's not exactly who I thought I would be when I was a girl," she says. She had been in the dark about her gifts for much of her young life, until she met Terry. Her story invites the question: What made her so strong?

I counter with different questions: How do we know others don't have that potential? And what if the answer is more banal than you think?

Before Jan emerged as a strongwoman, women were rarely allowed the opportunity to reveal themselves as such. "Even if somebody is born with a particular talent," observed the historian Yuval Noah Harari in his book *Sapiens: A Brief History of Humankind*, "that talent will usually remain latent if it is not fostered, honed, and exercised." Harari explains that we have a tendency to create social hierarchies based on imagined criteria. "Whether or not they have such an opportunity," he wrote, "will usually depend on their place within their society's imagined hierarchy." The reality is that women are just beginning to catch up to their potential.

"There was a kind of responsibility, after Title IX happened, where I was suddenly seen as a role model for others, to expand the concept of sport for women," Jan says. "There are women now who are professional strength athletes, who are deadlifting well over six hundred pounds. There's a woman who has squatted eight hundred pounds. A number of women have picked up the Dinnie Stones, and a woman picked up the Húsafell Stone"—a basalt lifting stone weighing 410 pounds, which resides in Iceland—"and walked with it. Nobody would have believed we would see a woman like that on this planet. It used to be that we thought that women were half as strong as men; then it was two-thirds. There are remarkable things happening all the time. If you look at history, guys have been thinking about how to be strong, training and competing, in military and sport capacities, for hundreds of years. For women, it's only been a few decades. We're still figuring out what's possible." Weightlifting records are constantly being broken, by men *and* women, with bodies of all shapes and sizes.

Back in 2019, twenty-three-year-old Chloe Brennan, a scrappy 140-pound nurse practitioner from Birchmoor, England, was working full-time and training on the side when she won the title of the UK's Strongest Woman. That same year, she'd also managed to lift the Dinnie Stones in Scotland, but it didn't go as well as she'd hoped—she wanted to make a better showing with a longer hold and a higher clearance.

Three years later, Jan was codirecting the 2022 Arnold Strongman Classic, in Columbus, Ohio, where an event featuring a replica set of Dinnie Stones would be one of the most anticipated events of the weekend. (In 2023, the women's division was officially renamed the Arnold Strongwoman Classic, to be parallel to the Strongman.) After Brennan missed a qualifier, she wrote to Jan asking for the chance to compete.

Jan knew that Brennan was capable. She said yes.

On the day of the contest, the air in the arena was electric. The crowd murmured and then quieted as two hopeful champions stepped up to the stones and failed to lift them. It's easy to view this as a scene out of time, one that has happened over and over again across human history, but now the contestants included women. Brennan had just thirty minutes of rest after the amateur strongwoman competition, in which she placed second in the lightweight class. She came back onstage in a kilt, ponytail swinging, breathing hard, to make her attempt. When I watch the footage, I'm struck first by how tiny she is—the Dinnie Stones are more than *five times her body weight*—and by how fiercely she approaches the lift, stalking up to the rocks like a predator.

She hauls the stones up off the ground: *One Mississippi, two Mississippi.* The sea of giants around her begins to roar.

When Jan tells me what it was like for her to watch Brennan, she speaks with the gravitas of a sports historian, but also with the awe and excitement of a spectator.

"There's a photograph of me standing on the edge of the stage, and I just have my mouth open," she says. "Chloe got the biggest cheer of anyone at the Strongman that weekend. What it means is that the world has changed. It's finally okay for women to be strong. It validated that women of all sizes can do so many things if they train."

She calls out the countless men over the years who sneered at women who had lifted the Dinnie Stones, saying that they weren't "real" women. "To watch someone like Chloe," she says, "amazingly strong, so full of joy, doing something that most men wouldn't be able to do?"

She can't help but laugh. "I'm all in on that."

THE WEIGHT BENCH in my father's art studio was surrounded by dumbbells and stacks of iron plates, to load onto a barbell for squats or bench presses from an overhead rack. My father lifted these weights regularly and taught us how to use them correctly. But this wasn't the central part of his strength regimen. Above all else, I remember the body-weight exercises, endless repetitions of them: the push-ups and the pull-ups, the planks and the handstands and the bridges. The body itself was burden enough. Learning all the ways to lift it was practical; repeat enough times, and you didn't need anything else.

Every push-up was a note to yourself—that you were capable of more than you were the day before. These days, when so much of life feels out of our control, something that concrete and tangible, however incremental, feels like a small victory. During the writing of this book, I've been doing a lot of push-ups, and thinking about what it means to do them. Lately, I've come to understand that what I learned from my father was not just to lift heavy things, but to lift myself.

FORM

Who has not been, or is not a good master of the figure,
and especially of anatomy, cannot understand it.

—MICHELANGELO

5
—

The Ideal Body

A muscle, once it is free from life's electricity, is suddenly limp, at a loss. Then it grows recalcitrant.

It stiffens, held in a state of contraction called rigor mortis. A few hours after a person or animal dies, the loss of energy molecules necessary for muscle fibers to relax causes a body's joints to become fixed in place. This process is generally temporary, but in the case of a donor body being preserved for anatomical dissection, a combination of rigor mortis and formalin can lock limbs in odd positions: a foot pointed like a ballerina's, an arm bent. It gives the impression of a body paused midaction.

The same stuff that activates us in life arrests us in death.

When observing a dissection, one cannot help but think existentially about these physical forms we inhabit. They are temporary vehicles, ones that we will eventually leave behind. The engine has died, but the vehicle is still parked in the lot, awaiting examination, should we have the privilege to peer under the hood.

"For someone to say 'You can take my body and learn from it. Here's my brain, here's my heart, here's my everything'—it's an honor," says Amber Fitzsimmons, a professor and chair of the Department of Physical Therapy and Rehabilitation Science at the University of California, San Francisco, with a joint appointment in the School of Medicine's Department of Anatomy. She leads me into the anatomy lab, on the thirteenth floor of the Medical Sciences Building, where you can take in what is arguably the best view in all of San Francisco. It's a place to admire the sweep of the city, across a continuous wall of windows, from the Pacific Ocean to the downtown skyscrapers, the plume of Karl the Fog rolling in across San Francisco Bay and reaching its tendrils east toward the hills. (Yes, the fog here has a name, and a very distinctive personality on social media.) It's an extraordinary piece of real estate for a room full of dead people who can't admire the view, but somehow the loftiness of the location is fitting.

Two medical students are performing a dissection at the far end of the room, and Amber greets them warmly.

"To see what's underneath, it can be shocking. It *is* shocking," Amber says, turning to me.

For this reason, there is a process in place for human donor dissection, and with that process comes a reverence that helps you to understand the privilege of getting to look. Head, hands, and feet are wrapped before dissection. And at the end of every academic year, a special memorial is held, so that the school's medical, physical therapy, dental, and pharmacy students can honor those who essentially served as their very first patients. The donors' ashes are then scattered at sea.

If you're wondering why I want to witness a human cadaver dissection of my own free will, I'll be honest with you—I've wondered about that too. As someone who has had a deep fear of death from a young age, I don't take the enterprise lightly. But I suppose I seek it out because I want to understand—in a way that's essential, firsthand, and, yes, *visceral*—what lies under the surface assumptions we have about our appearances, and how muscle shapes the way we see ourselves. I'm visiting an anatomist because I want to actually *see* muscle for the first time. Muscle forms the way we look; we ascribe aesthetic value to the appearance and shape of the body. How we view muscle in turn influences how we view a person in the world.

If beauty seems far from the point, remember that beauty was once a primary focal point of dissection. Michelangelo Buonarroti and Leonardo Da Vinci, two of the greatest Renaissance artists of the human form, both felt that a scientific understanding of the body was necessary to portray it. The practice they had of observing and performing anatomical dissection undergirds our modern understanding of ideal beauty.

I want to get back to that source material: the muscle itself. What does looking at it tell us about appearance, function, and what we value? What makes a muscle beautiful?

As we approach our own table, a white cloth covering the body, Amber asks, "Ready?" and I nod. She pulls back the sheet, and readies her blade for the first cut.

ATLASES OF THE body recur across civilizations and cultures. Muscles are so front and center to the Western understanding of the workings of the human body that I am startled to learn that muscles did not figure into traditional Chinese medical maps of the body.

"Chinese doctors lacked even a specific word for 'muscle,'"

noted the science historian Shigehisa Kuriyama in the seminal book *The Expressiveness of the Body and the Divergence of Greek and Chinese Medicine*. He illustrates the point by comparing two classic and influential depictions of the human body: a series of full-body muscle plates for Andreas Vesalius's *On the Fabric of the Human Body in Seven Books*, published in the sixteenth century, and Hua Shou's acupuncture diagram for *Routes of the Fourteen Meridians and Their Functions*, from the fourteenth century. To Mongol and Ming dynasty–era Chinese doctors, muscles were essentially invisible; the chief preoccupation when it came to physical health was managing the flow of energy within the body.

The ideal body in ancient China was one that was full of vital breath, or *qi*, and was self-contained. A strong, powerful, and healthy body did not leak qi. It wanted to exert itself, but the exercises that were prescribed for this body focused on stimulating flow and flexibility, not on creating an articulated, muscular physique. It did not look like Hercules but, instead, resembled a yogi with a full belly. That belly might look unfit to us today, but in ancient China it was not—it was full of qi.

Like their Chinese counterparts, ancient Greek physicians in the time of Hippocrates, circa the fifth century BCE, did not generally distinguish muscles from flesh overall. But Greek artists of that period *did* sculpt their human figures with bulging ripples, even in places where no muscles existed. The appearance of an articulated body was deemed attractive, even before the Greeks understood what muscles actually were. By the time of Galen, the influential second-century physician and philosopher, the rise of anatomy and dissection had encouraged a new perception of muscularity and agency that was tied together.

In his writings, Galen described muscles as the drivers of voluntary action, and thus our human selves. Kuriyama explains that the ancient Greek preoccupation with muscles is "inextricably intertwined with the emergence of a particular conception of personhood." Muscles were not only about external ideals, but about internal ones too. When we talk about muscles, what we're really talking about is the expression of *who we are*.

It's a philosophy of life that still very much animates Western thinking today: Our actions reveal our character, the choices that we make. And our muscles are the agents of that autonomy, over our bodies and our lives.

THE YEAR I turned twelve, 1989, coincided with the Vatican's completed restoration of Michelangelo's Sistine Chapel ceiling to its magnificent original sixteenth-century coloration and clarity. My father marked the occasion with a foundational lesson in what it was to portray the human body in the robust glow of life.

We examined close-up details of various biblical characters in art books and magazines. The ferocious physicality of the sybils and the prophets struck me; I loved the androgynous beauty and power in their forms.

Look at those muscles. My father pointed out the radiance of the figures, the solidity behind all the light.

One art restorer, who had spent years carefully cleaning away centuries of grime from Michelangelo's vividly rendered figures, remarked that an observer could "see the blood flowing through their veins."

Michelangelo's intimate knowledge of human anatomy had come from scientific dissection, the practice of which had been revived in medical schools in university centers like Bologna and Padua but was still highly restricted by the Catholic Church in Renaissance Italy. Artists couldn't easily obtain bodies to dissect, but religious authorities gave Michelangelo more leeway than most—he elevated the portrayal of the human body to something approaching the ecstatic. The skillful depiction of biblical stories in church frescoes was critical to attracting the general populace, most of whom could not read. A beautiful work showing the lives of the saints was considered essential religious education.

The Sistine Chapel ceiling took Michelangelo four years to complete; it was unveiled in 1512. A quarter of a century later, at the age of sixty-one, Michelangelo was once again drafted into service by the Vatican, to cover the grand wall above the altar with a final fresco: *The Last Judgment*. It was the source of both awe and scandal—one writer described it as "a tumultuous sea of jumbled bodies," with Christ depicted as a powerful Apollo-like figure and wingless angels as muscular enforcers beating the damned back down to hell. The nudity and anatomical specificity of the figures shocked the public, but even then the palpable *humanness* of the work was considered an astounding feat.

The Last Judgment, it must be noted, was also a strikingly physical act of endurance for the aging artist himself, who climbed seven levels of scaffolding each day for five years.

He even wrote a satirical sonnet about how miserable it was to paint the ceiling. "A goiter it seems I got from this backward craning . . . ," he wrote. "From all this straining / my guts and my hambones tangle, pretty near." In the margins of the poem, he sketched a bent figure contorting itself to paint shapes on the ceiling.

Toward the end of the project, Michelangelo fell and suffered a serious injury to his leg. But he kept at it, and saw the fresco to completion.

ART WAS, LIKE exercise, the connective tissue between my father and me. He was an artist who won an Emmy for his work in 1977, the year I was born. For twenty years, he did mostly commercial work, drawing and painting illustrations for film posters, print advertisements, romance novels, and children's books. When my father made his regular train trips from our home on Long Island into Manhattan to see clients, I often went with him. We'd end up doodling together—in the company of French realist painters like Bouguereau or Bastien-Lepage—at the Metropolitan Museum of Art.

When I said I wanted to be a freelance artist like him, he used to tell me I should be a plastic surgeon instead. "That way you can still play with people's faces, but you'll make a lot of money!" he'd say. It was a solution he found genius but I found gruesome.

Only recently did I learn that my grandfather had worked at a mortuary in Kowloon for many years; as teenagers, my dad and my uncle Peter lived with him in municipal housing above the morgue. I asked my father if they ever went inside. "Oh, yes!" he exclaimed. "We hung out there. There were body parts in jars." Looking back, it's clear that even in childhood the connection between art and anatomy had been made, for both of us.

I don't remember when I learned to draw. By kindergarten, I knew what a maulstick was, and stick figures were not in my repertory. This was part of the education I acquired from my father, among other things I inherited, like the shape of my feet and my quirky, maneuverable eyebrows.

In the late 1980s, he started doing covers for the Choose Your

Own Adventure series, which at the time were among the most popular children's books in the world.

You have farmed with your uncle in rural seventh-century China for as long as you can remember. Your long, tiring days in the fields come to an end when barbarians raid your small village. Will they torment you as a prisoner, or take you to a more exciting life?

The wide-eyed peasant boy on his knees in supplication on the cover of Choose Your Own Adventure #109: *Chinese Dragons*? That's my brother. That imperious warlord all armored up before him? That's my dad.

Do you stay in the field as your uncle orders? Or do you run away to follow the soldier-hero Li Shi-min, to train and become one of his young warriors?

We followed our father faithfully. Just as he trained us in the studio to punch and kick like him, he also trained us to draw and paint like him. Shaping the body in art and life went hand in hand. They were methods for exploration: to play, to shape-shift, to see who and what else you could be.

When I was in high school, I painted myself into the Sistine Chapel ceiling, as part of an art class project on self-portraiture and the Italian Renaissance. The assignment was to create a self-portrait in the style of an old master. I picked Michelangelo because my father loved Michelangelo, and because the frescoes meant something to me. My father was largely gone by that time, living as he was in Hong Kong most of the year. My parents had begun a separation but had never named it as such, and I suspected that he had a whole other life that he didn't tell us about.

"Maybe he has another family," my best friend mused aloud, and I felt my hackles rise.

The experience of childhood is suffused with a peculiar feeling of fragmentation, when you only get part of the picture, and you struggle to understand everything from the pieces you have. You're afraid to ask. Truth be told, you don't even know *what* to ask.

In my father's absence, I took up residence in his studio. If he wasn't around, I'd at the very least use his paints, for a project of my own reinvention.

Michelangelo was subversive in many ways, not least because he painted all the women of the chapel with the same robust muscularity as the men, with no distinction for gender. The patriarchal Church relegated women to the home, and obedience and modesty were valued most of all. Ignorant as I was at the time of the religious underpinnings, I still liked the muscles. Michelangelo evoked the body, with its barely contained power, as a volcanic expression of the soul. (This is also not a bad description of the teenage state of being in general.) I painted myself as the prophet Daniel, seated on a marble throne and deep in concentration over an open book; the book was borne up in turn by a smaller figure, who shouldered this burden of knowledge like Atlas holding up the world.

Over several weeks, I sat in front of a wall mirror in the studio and painted this life-size version of myself as proud, muscular, and unrepentantly strong. The painting still lives in my mother's house today.

Biblical scholars say that Daniel was an interpreter of dreams, the hero of his own book. On the ceiling, Daniel peers to the side, appearing to make a notation with his right hand; it's in this hand that I placed a paintbrush. In my rendition, though, I look directly at the viewer, calm and confident. This was my story, and I was beginning to tell it. Despite all the hormones and high feelings of sixteen, this version of me didn't seem far-fetched at all.

6

Who's Afraid of a Lady Hercules?

A muscular woman has historically been a difficult woman. The way we perceive beauty in the muscular form is influenced by many factors, not least of which is gender. The difficulty, I've come to understand, is rooted in disruption. Through the ages, muscular women have "disrupt[ed] the equation of men with strength and women with weakness that underpins gender roles and power relations," observed Patricia Vertinsky in *Venus with Biceps*, a history of strongwomen that explores femininity and muscularity through a wide-ranging collection of vintage posters, comic books, advertisements, photographs, and other ephemera from 1800 to 1980.

These images reveal our deeply conflicted relationship with strong female bodies, over a period of nearly two centuries—when circus performers, trapeze artists, gymnasts, bodybuilders, and other athletes could be greeted variously as (a) daring and fascinating, (b) exotic and titillating, or (c) disturbing and morally repugnant.

Binary thinking about gender roles has long encouraged a worldview in which muscularity and physical strength for women is a threat to societal order, domestic harmony, and reproductive responsibility. Societal concern for the "defeminizing" impact of sports was once widespread. Sports have been deemed a road to moral ruin for women by nineteenth-century Christian authorities in the United States and lawmakers of twenty-first century Saudi Arabia. Well into the twentieth century, women were warned not to lift weights or other heavy things, or their uteruses might fall out. These days, a woman with muscles is more likely to be seen in a positive light, but even now there are caveats.

The phrase *too muscular* lives in this vast cultural minefield. It might seem to be a superficial matter of aesthetics, but it is a charge that has been levied in many arenas against women who dare to be physically strong. Jan Todd, for one, disrupted stone-lifting practices that were for centuries linked to rituals of manhood; after all, what significance did manhood stones possess if women could lift them too?

Too muscular can call into question one's identity as a woman: Are you a real woman if your muscles are bigger than the societal norm? *Too muscular* can be accusatory: Are you a cheat, guilty of using steroids or other performance-enhancing drugs?

The words can also signal veiled racism. At a 2017 talk with students at Harvard, the ballet superstar Misty Copeland spoke about close-minded ideals of beauty. "Why am I being told my body is too muscular?" she said. "It's code language for your skin is wrong." The tennis great Serena Williams, in a 2016 interview with the *Guardian*, said that she has been described as "too muscly and too masculine, and then a week later too racy and too sexy." In white-dominated spaces like ballet and tennis, *too muscular* can be code-speak for "too Black," for bodies that don't belong—often jumbling up issues of femininity, race, and power.

Too muscular is also used to disparage transgender female and intersex athletes with naturally high levels of testosterone. The growing controversy over the participation of transgender women in athletic competition is rooted in muscle, and the perceived unfairness of muscles that come with male puberty. This, of course, disrupts the long-standing division of sports participation based on sex. The recent establishment of nonbinary divisions for major marathons including the Boston Marathon and the New York City Marathon is one way that organizers of athletic competitions are addressing the issue. There will undoubtedly be more rethinking to come.

Strength, of course, is not zero-sum—if I am strong, you can be strong too. But zero-sum thinking has undue influence over the idea of feminine power. When a woman is deemed too muscular, it's often because her strength is perceived as taking away from someone else, or that her strength is somehow unseemly, unfair, or unnatural. All kinds of wacky theories around hormones have been used to delegitimize women in power, connecting the body to the body politic: menopause has been called out as something that makes women unstable leaders (see: Clinton, Hillary), and yet testosterone is the hormone that actually makes people reckless (see: Clinton, Bill).

All this is to say that pseudoscience has long governed norms around women's anatomy and biology. Maybe our viewing habits around muscular beauty have gotten a bit entrenched. If we go back to the muscle itself, could that shake up our thinking?

AMBER FITZSIMMONS IS a modern-day anatomist—a professor of anatomy who instructs students at one of the top medical schools in the country, as well as a physical therapist who has seen all kinds of real, actual bodies enter the clinic. During my visit to the UCSF anatomy lab, I ask what *too muscular* means to her, and she reminds me that Americans have been socialized to not want to see the woman weightlifter body, the bulked-up form that became especially taboo in the 1970s and '80s.

I recall the East German and Soviet sports machines, with their state-sponsored doping programs designed to gain political and cultural currency in Olympic competition during the Cold War; the panic over these secret practices is what led to systematic "femininity control" and sex testing in sports.

"'Too muscular' means 'too masculine,'" Amber says matter-of-factly. "You don't want to be seen as a man. And that fear still persists around women and exercise."

Amber spends a lot of time thinking about what muscles mean. For her, form and function go hand in hand. Before our dissection, she gathered an audience of anatomist colleagues—Dana Rohde, Barbie Klein, and Maddie Norris, all instructors and researchers at UCSF—on my behalf, to help unpack the vocabulary of muscle, what makes a muscle beautiful, and where the gendering of those ideals comes from.

When I first encountered this group of anatomists, they were all standing around in the hallway telling anatomy jokes—which, to be honest, sounds like the setup for a joke in itself.

"What's the difference between the testes and the prostate?" Dana asked.

"There's a *vas deferens*!" (Cue uproarious laughter.)

We talked about muscles and their functionality, but also about what they present in a broader context across various spaces: language, culture, society. What kind of modern perspective on muscle could these anatomists offer?

"I'll be honest with you, we do have favorites," Amber said with a grin.

At fifty-two, she is a lifelong swimmer, with good posture and an easy laugh. Her capacious enthusiasm for all things muscle invites questions—even outlandish ones—because she considers each one with thoughtful earnestness.

Swimmers' shoulders. Runners' legs. Gymnasts' abs. Dancers' posture. These phrases summon up different body types, all admired in one circle or another. (I admit that shapely soccer-player calves drew me to my husband when we first met.) They raise the question of what is behind the appeal of specific muscles and the characteristics they connote.

Look at the way muscle insinuates itself into the lexicon. Synonyms for *muscle* include *potency* and *domination*. When you force someone to agree with you, you're *strong-arming* them. To *make a muscle*, you contract your biceps—or more accurately, the biceps brachii. There is no more stereotypical symbol of strength than the bent-arm curl—in fact, it's the stand-in for *all* muscle (see: emoji). And yet, despite its visible prominence, Amber explains, the biceps is the strongest arm muscle only when the arm is in this "Popeye" position—otherwise, it's the brachialis, a deeper, "pure flexor" muscle, which generates the most force, relegating the biceps to a supporting-player role.

What does a person who studies and teaches anatomy think when they see muscles on display? The room started buzzing with debate.

"Well, if you look at bodybuilders," Dana said, "sometimes their muscles are all for show—all that bulk makes it difficult for them to walk, and their lats are too big for a natural arm swing."

Contrast this with gymnasts, Barbie pointed out: "They can

lift their whole bodies with their hands, with such control—for me, what makes a muscle beautiful always goes back to function."

I thought about Marvel superheroes. Are *their* muscles functional? When my brother and I were kids, our father gave us comic books—X-Men, Wolverine, Dark Phoenix—to motivate us to draw human anatomy. We were instructed to study superhero physiques and practice sketching. What I absorbed from those comic books—other than the multiverse of stories, which I loved—was that male superheroes were top-heavy with biceps and that female superheroes were top-heavy with boobs. And that drawing that fictive landscape of muscles was a lesson in the American cultural psyche, with impossible ideals.

I wondered aloud: "What if you were a Hollywood trainer for a superhero movie? What specific muscles would you target to give the appearance of strength, on the ideal body, to an American audience?"

"Let's start with a quintessential male superhero: Captain America," Amber said. "Certainly, the arms—triceps, biceps. Then deltoids, pectorals, and latissimus dorsi, to create the exaggerated triangle from wide shoulders to a narrow waist. They overbuild the upper trapezius—that's around the neck—for a wide shoulder, then define the thorax with the external obliques"—the most superficial of the lateral abdominal muscles.

And, finally, the rectus abdominis—the six-pack.

"It's funny that if we see someone with a six-pack, we automatically think they're strong and really fit," Barbie added thoughtfully. "But they might just be naturally leaner."

Our discomfort with muscles begins when we move too far into that same territory for a woman. "The female equivalent is not equivalent at all," Amber said. "Female superheroes are strong, but they'll have boobs and a bottom. Smaller shoulders—not too wide. You'll have a flat stomach, but you won't see a supercut six-pack. Enhanced hips and glutes, tapering to a narrow waist—a controlled hourglass. You can't be too *extra*. If you see the thick neck, thighs, and wide shoulders that we expect on a man, it throws people off—and that's because we've been conditioned that way." In other words, we allow a greater spectrum of muscular beauty for men—from the lean, wiry marathon body to the big, beefcake muscles of the heavyweight wrestling body.

Even among female athletes themselves, there is a self-perceived conflict between their "performance body" in the sports context and their "appearance body" in the social world—across multiple studies of NCAA athletes in different sports, women have expressed pride in the utility of their muscularity on the playing field, but also worry that those same muscles would make wearing jeans or dresses look "abnormal"; they compensate by holding back in the weight room to avoid getting "too big" and by wearing makeup to emphasize their femininity.

When it comes to the superhero body, it's all about *signaling* fitness and outward muscular appeal rather than actual function, no matter what the gender. Theirs are the muscles that we—the audience—are indoctrinated to receive. We absorb that information into our daily lives and respond in kind. "All you need to do is go to a gym and see what's happening there," Amber said. "It trickles down."

THIS KIND OF thinking, it turns out, isn't just Marvel comics, Hollywood superficiality, and gym culture talking—it's embedded in our medical textbooks too. During one of our first conversations, Amber showed me the latest edition of the *Netter Atlas of Human Anatomy*.

"Frank Netter is a premier illustrator for the anatomy textbooks that many institutions use today," she explained, paging through the atlas, which was first published in 1989. "This is what the anatomy books generally show: white skin, perfect musculature, no body fat, no different skin tones, very gendered. I start my first-year anatomy classes by exposing that—the false perfection of a body as white and masculine, based off one specimen of surface anatomy."

I will point out here that a deeply troubling vein of white supremacy runs through the history of anatomical atlases: During Hitler's Third Reich, an unquestionably brilliant Viennese anatomy professor named Eduard Pernkopf employed an army of fine artists to create a four-volume atlas of human anatomy. *Pernkopf's Atlas of Anatomy* is a detailed work of scientific observation and illustration, but it relied on the bodies of those murdered by the Nazis, possibly including victims of concentration camps. It is still in use today, and the ethics of teaching from this book continue to be widely debated.

In the Netter book, we examined an illustration of a leg with exceptional muscle definition. "It looks like a Michelangelo marble," Amber said. "The reality is that hardly anyone looks like that."

In this way, muscle iconography in modern society can be harmful to men too. The social psychologist Jaclyn A. Siegel has studied how the stereotypical male body ideal contributes to eating and muscle dysmorphic disorders. In the attempt to become muscular, she has said, men are vulnerable to "the masculine norms of dominance, confidence, sexual success, and physical and emotional self-control," which make them susceptible to eating disorders. In fact, the quiet increase of boys and men seeking help for disordered eating, excessive exercise, and performance-enhancing substance abuse reveals how surface ideals of muscularity can hurt us all.

The signaling starts early, in everything from superhero movies to social media. A recent headline from a normally quite staid Harvard Medical School publication sounded this alarm: YOUR FIVE-YEAR-OLD BOY SHOULD NOT CARE ABOUT SIX-PACK ABS.

Long-standing biases can be difficult to see if you've never been taught to look for them. But little shifts are happening all the time. Norms vary by culture and geography, and they aren't static. Medical textbooks are beginning to feature more varied bodies; influential athletes are becoming more visible and vocal about body image and mental health; in the neighborhood gym near me in Berkeley, California, you'll find just as many young women powerlifting at lunchtime as men.

"I see and understand these biases as a human in society, and as someone who peels back the layers on what's underneath," Amber tells me, about how she tries to teach her students to observe more critically. "And as I get older, I'm less afraid to challenge the status quo."

IN 2019, ITALIAN art restorers were faced with a strange problem. The decomposing corpse of a murdered Medici duke, improperly embalmed, was eating away at four of Michelangelo's most famous muscular marbles: the allegorical *Day*, *Night*, *Dusk*, and *Dawn* sculptures in the Medici Chapel mausoleum complex in Florence, Italy. Over the centuries, phosphates and other material from Duke Alessandro's body had infiltrated the luminous Carrara marble, leaving it pockmarked and deeply discolored.

Restorers had to carefully deploy hungry microbes to hoover up Alessandro's remaining phosphates and clean up centuries of stains. The irony that even a perfectly sculpted representation of ideal beauty could be marred by the bodily processes of an actual decaying human tickles me. Here was a body long past its prime, reminding us that, despite our best efforts, time will always humble us and keep us real.

7

Shoulders, Squared

"There are few body parts quite as freighted with symbolism as the shoulder," wrote the fashion critic Vanessa Friedman. "Squared, they take on responsibility and life's burdens. Bowed, they indicate humility, pain, fear, reverence. Shrugged, they signal indifference. Built up, superheroes, super villains, and the superglam. They are a resting place for angels and the weight of the world."

Friedman's observations about shoulders as body part and cultural signifier extend beyond what we wear. Our shoulders are more than a coat-hanger frame; they are capable of communicating feeling and apathy, power and weakness. Shoulders carry shades of meaning physical and existential to us; many other muscles do the same.

To me, the shoulder is one of the most seductive and beautiful muscle groups, but its external appearance is just one reason. The metaphorical quality of the shoulder also appeals. *To shoulder something* is an act of generosity, especially in service to another. As it pertains to all that a donor body can teach a student of medicine, the thematic circularity feels especially apt.

For all these reasons, I'd asked Amber to dissect the shoulder complex for me, and she happily accepted and argued the case for both form and function. (Full disclosure: as swimmers, we are both biased toward a really good-looking set of shoulders.)

"Nothing works singularly in the shoulder," she says now as she began to cut and peel away the skin and fat from the back of the shoulder. There, on the table: meat and sinew. Neither of us can look away.

Above her surgical mask and scrubs, Amber's blue eyes widen in emphasis; in a bit of flair, a colorful tie-dyed bandana holds her hair back from her face as she works. I, too, wear a mask and scrubs, but I largely keep my gloved hands in my notebook.

"When people talk about the shoulder, they mostly refer to the deltoid, but there are seventeen muscles that act on the scapula," she explains, pointing to what most of us know as the shoulder blade. "There's all this coupling that's required to be dynamic in a wide swath of action, and to do so with precision. And when you look at how the shoulder is connected to the thorax and the sternum, with all these muscles and tiny ligaments—well, it's gorgeous."

I wasn't sure how I'd feel in the presence of a dead body, but once Amber starts talking, I find myself calmed by her steady stream of enthusiastic commentary. At the most basic level, anatomy is the study of structure: how a body is organized, how it works. Separating a body into its parts reveals how the whole moves together.

As Amber manipulates the joint to soften the preserved tissue, she ruminates aloud, "I think the deltoid is the most boring of the seventeen, because it's so *obvious*."

When I laugh and accuse her of playing favorites, she laughs too. "Well, the deltoid actually has three parts, and it does act as a flexor, an extensor, and an abductor," she says—which means it bends, extends, and moves the arm away from the body. "But I don't find the movements all that exciting. People like the deltoid because they like the way it looks. It's superficial—it's the one people see, and it's also easy to build up. And it's the deeper muscles—literally deeper, under the surface, and also figuratively, in that they're more complex—that are more dynamic."

She traces the line of the serratus anterior, the fan-shaped side muscle that goes onto the thorax and slides underneath the scapula. "The line where the serratus goes—now this one is special," she says. "It's such an important muscle for arm function—when you raise your arm up, there's a dance that happens between your scapula and your arm, and the serratus couples with the trapezius to make that happen."

When something goes wrong in the shoulder, she adds, a lot of basic function is put at risk. The muscles at work in the shoulder are behind so many of the movements that are essential to everyday life: brushing your teeth, putting your shoes on, sticking your phone in your back pocket. It's incredibly intricate, and it's also what makes it fascinating.

She casts about for the right analogy, and her eyes light up. "It's like an orchestra, every day."

Once Amber has painstakingly removed the skin and fat to reveal the pearlescent, membranous layer of fascia that holds each muscle in place and in its discrete shape—"I always tell my students that every muscle is a letter, and the fascia is the envelope"—I begin to see the inner workings in all their complexity: how the muscles of the shoulder slide so beautifully over one another, how they attach to bones at different points to support complete, all-around movement, how they have to work in concert. I can hear the orchestra tuning up, and I can also appreciate the distinct sound of each individual player. We admire the pennation: the feathery quality of the muscle, the fibers of the serratus fanning outward from the attachment on the bone.

Anatomical dissection was once an event that was open to the general public. On the frontispiece of the original sixteenth-century edition of Vesalius's meticulous *On the Fabric of the Human Body*, there is a wood-block illustration of Vesalius himself—considered the founder of modern anatomy—dissecting a corpse in a dense, jam-packed amphitheater.

Several months earlier, I'd had the privilege of examining an original copy of Vesalius's work in the rare books reading room of the British Library in London. What struck me most about the dramatic peeling back of the layers was the audience rapt for knowledge, eager for secrets revealed. How do we work? The drawings attempt to answer this desire with fresh clarity. Every muscle in place, poised for purpose, ready to pull the levers of the human machine.

These days, anatomical dissection is decidedly less public—a rarity, really. Most people will never encounter an experience like this, and I come away enamored with the stories that a body can tell. The awe and opportunity of looking inside a body means that you will never again look at the outside of it in the same way.

When it comes to our donor, I can't tell you too many details for privacy reasons. Amber takes care to point out that there are no hard and fast rules about muscle on a younger person versus an older person, regardless of sex; no matter what the profile, you might find that the muscle is in perfect health, its fibers distinct and easy to examine, or that fat has infiltrated it and marbled it with age or disease. Our donor, despite their advanced age, had well-defined muscles, and their tidy structural integrity brings Amber back to the awe she felt when first learning about the human body. "Like art coming to life," she says.

I love that the experience of dissection encourages us to rethink long-held assumptions about the external appearance of muscles and the body. You can never predict what you're going to see until you've gone in. And muscles hold fascinating secrets that only they can reveal.

The reality is that muscles are always changing character. "How we use a muscle [and] how the central nervous system sends signals to it determine how it will develop, not just in size but in structure and biochemical properties," wrote biologist Steven Vogel. He noted that though all muscle is pretty much the same at birth—think of the pale meat of young mammals, like veal—skeletal muscles respond to training stimuli with long-term and short-term changes.

We have fast-twitch and slow-twitch muscle fibers. In general, muscles with a lot of fast-twitch fibers are ideal for power movements and quick contractions over a short period of time—like the hamstrings, located at the back of the thigh, which are used for sprinting. These are also known as white muscles, or white meat—yep, just like poultry—because they appear lighter in color. Fast-twitch muscles get their fuel from glycogen—the stored form of glucose, for when you need quick energy—and require less oxygen, so they have fewer blood vessels and mitochondria. By contrast, muscles rich in slow-twitch fibers are good for endurance—like the soleus, located in the calf, a key muscle used for standing. These are known as red muscles, or dark meat, and they need lots of oxygen to fuel long periods of movement;

they are rich in blood vessels, mitochondria, and myoglobin, the oxygen-binding protein in blood that gives muscles their reddish color. (On average, women have a larger percentage of slow-twitch fibers, which is one reason that women are more fatigue-resistant than men after ultra-endurance activity.)

Shoulder muscles are roughly half fast-twitch and half slow-twitch fibers; the muscles Amber and I are examining in dissection appear grayish red, but it is difficult to tell their original color, since the donor body was treated with formalin and the chemical in the vascular system dissipates that coloration over time.

Though we're all born with a particular number of muscle fibers, which largely dictates our muscular potential, we are not completely predestined. A study of identical twins, one of whom ran marathons and did triathlons, and one who stopped due to an injury in high school, revealed that by the time the two brothers had reached their fifties, the endurance athlete was 94 percent slow-twitch muscle fiber, while the sedentary truck-driving twin was just 40 percent. Despite starting from the same genetic place, phenotypic plasticity allows us to change our muscular makeup to a remarkable degree.

Skeletal muscle can even be transformed into *heart muscle*, after a fashion: In a procedure called dynamic cardiomyoplasty, a piece of fast-twitch latissimus dorsi can be wrapped around a ventricle weakened by heart disease and convinced to turn into slow-twitch muscle in just a couple of months, with the help of electrodes that approximate the heart's own normal stimulating signals. You could say that muscles are open to possibility.

We think a lot about our brains, as kind of the control rooms steering these ships we call our bodies. Ranking significantly lower on our priority-organ list is muscle—in fact, we've long thought of muscle as rather dumb. (Muscles are something we continue to have unresolved anxieties about. Though we mythologize athletes, we obviously have contradictory feelings about muscle taken to the extreme.) On the whole, people with prominent musculature are often saddled with the assumption that they aren't smart; "dumb jocks" are believed to have more brawn and beauty than brains. We perceive muscles as taking away from intelligence, when they actually have their own kind of intelligence.

Muscles are smarter than we think. They have different personalities. They remember things. When they change and grow, they influence other body systems to do the same. They're *complicated*.

MORE AND MORE, I notice the ways that people attribute character to muscle. What might a body atlas correlating muscles with personality traits look like?

Serratus anterior: gets along well with others. "Without it, your arm function is pretty much a goner," Amber tells me. "And who doesn't love to see the gorgeous outline of the serratus interlocking with the external abdominal oblique for a truly magnificent piece of art?" (Remember, this is a swimmer talking.)

Teres major: petite but very strong. Dana, who teaches musculoskeletal anatomy to all first-year UCSF medical students, has a special interest in climbing and yoga. "As a climber and yogi looking at a lot of backs, I just think it's a really pretty muscle," she says. "Maybe also because it's very small, and it sneaks in to join the lats. Petite but strong—like me."

Piriformis: infamous for being a troublemaker. Maddie tells me that this little pear-shaped muscle in the hip is involved in all kinds of reasons people, including herself, go to see a doctor, including sciatica, low back pain, and numbness in the butt and the back of the leg. (Maddie also notes with tongue-in-cheek humor that, just like the "troublemaking" piriformis, it is in her own character to quietly complicate things, by providing counterarguments.)

Quadriceps: strong and powerful. Every so often, the quads point the way to becoming someone else.

In the early 1960s, the neurologist and writer Oliver Sacks lived for three years in an apartment near Muscle Beach in Venice, California. Early on, he lifted 575 pounds in an informal head-to-head contest, and earned his membership in this famous circus of muscular misfits.

"I was accepted on Muscle Beach," he wrote, "and given the nickname Dr. Squat." In 1961, he set a California state record in weightlifting with a full squat of 600 pounds.

In the memoir *On the Move*, published shortly before his death in 2015, Sacks reflected on the reasons that he lifted weights for so many years with such focused, relentless intensity: "My motive, I think, was not an uncommon one; I was not the ninety-eight-pound weakling of bodybuilding advertisements, but I was timid, diffident, insecure, submissive." Muscles were a stand-in for the character traits that he desperately wanted to have instead.

He wished to change something essential in himself through muscle. But forcing his form to change so radically, pushing his muscles "far beyond their natural limits," came with a price. The years passed; he ruptured one quadriceps tendon in 1974, then the other, a decade later. And despite all that he had accomplished, he still lacked the self-acceptance that he was after most of all.

"While I was in hospital in 1984, feeling sorry for myself, with a long cast on my leg, I had a visit from Dave Sheppard, mighty Dave, from Muscle Beach days," Sacks wrote. "He hobbled into my room slowly and painfully; he had very severe arthritis in both hips and was awaiting total hip replacements. We looked at each other, our bodies half-destroyed by lifting."

"What fools we were," Dave said.

Sacks just nodded.

We want what we don't have, or what we don't perceive ourselves to have. Sometimes, when we take things to the extreme, we hurt ourselves in our aspirations. The way we reshape our bodies speaks of this poignantly human desire to change what we don't like about ourselves.

I'm not immune to what the culture tells me I should look like. Sometimes I wish I could be described as tall and willowy, rather than a completely average five foot four. During pregnancy, when my bra size temporarily increased from an A cup to a C cup, I was amused by the novelty. But I like that sports helped me to not be ashamed of my body's appearance. I have my own vanity about my swimmer's shoulders, my strong legs and kick; I like what they tell me. To be proud of their power, of how well they work.

As a kid, I learned to do handstands from my dad; as an adult, I began doing headstands as a regular practice in yoga. If I ask myself why I still do them, I realize that I like going upside down because it not only encourages a radical shift in body awareness, but also in perspective.

"I've always thought that inversions make you aware of your muscles in different ways," says Barbie in the anatomy lab. She proceeds to tell me about the deep proprioceptive and postural muscles, including the suboccipital and cervical muscles of the neck and spine, and it pleases me to know the names of those muscles that are helping me find myself in space anew.

When I kick my legs up from the floor to stand on my head, these muscles will help me to lift, to look, to change. To be open to the world and its possibilities.

I MADE MY pilgrimage to the Sistine Chapel the summer after my sophomore year in college. My roommate Melissa and I worked for months to save up for that early-August trip to Rome, Florence, and Venice. My parents had finally divorced, and my father had moved from Hong Kong to Beijing. My mother had to sell the house I grew up in, and there was little money for anything beyond the necessities. But I knew, even then, that this trip was in some way essential to my very being.

A friend of mine, a gruff and talented artist who was two years ahead of me in high school—this alone lent him an air of cool authority—gave me a piece of advice: When you get to the Vatican and they open the doors for the day, run all the way through to the end of the galleries. You'll get ten minutes alone in the Sistine Chapel.

I was a twenty-year-old college kid, and I knew next to nothing about the world. The only international travel I'd done was summer visits to my paternal grandfather in Toronto, Canada, before he died prematurely, from that heart attack. I remember chatting dazedly with the customs agent upon landing at the airport in Rome—"*Non ho niente da dichiarare*" ("I have nothing to declare"), a phrase that comes easily to me even today—and then exchanging excited hugs with Melissa, our heavy backpacks swinging so wildly that we came close to knocking each other over.

We were thirsty for beauty. We stood in long lines at the Uffizi Galleries, where I passed the time making line drawings of the crowds in my sketchbook; we sat on the Ponte Vecchio and people-watched, shooing away catcallers roosting nearby on the bridge. One sunset evening, picnicking on a hill high above Florence, Melissa and I shared a bargain bottle of pinot grigio, then went in search of gelato for dinner. We roamed everywhere on foot and rode the overnight train to Venice in a stuffy car packed with restless families. We chased pigeons in Piazza San Marco and allowed ourselves to get hopelessly lost along Venice's meandering canals, made for digression.

When it came time to visit the Vatican, near the end of our trip, we followed my friend's advice. We lined up early at the museum entrance. When the doors opened, we took off, like sprinters at the starting gun.

We did the bob and weave, threading our way around the other tourists in the galleries, the throngs growing ever thinner. When we reached the Sistine Chapel, we caught our breath and looked at each other, then quietly stepped over the threshold, the first visitors of the day.

I remember walking to the center of the room and lying down on the floor. I remember staring up at that jumble of muscular bodies, the figures aglow with life, the blood running through their veins. I remember that I began to cry.

It could have been ten minutes or ten years. I cried because I was moved by beauty and awe. I cried because I had made it there, on my own steam. And I cried for the loss of the childhood relationship that I had loved having with my father. He was the reason I was here.

And yet: On the cusp of adulthood, I was becoming someone else.

SIX MONTHS LATER, during my junior year studying abroad in Sydney, Australia, I bought an international calling card and dialed my father's number in China. I told him I wanted to visit him on my way home; it had been three years since we'd seen each other, and he rarely called or wrote.

"It's not a good time," he said, his voice crackly, a little tinny over the oceanic divide. "Maybe next year."

I yelled and cried into the receiver. I told him that if he didn't let me go see him right then, I would never agree to see him again. Who was this college kid issuing ultimatums? I was constitutionally averse to conflict, always had been. But our estrangement had made something bitter boil up inside me.

I went to Beijing that year because I wanted to begin a process of repair, though I didn't know what that might look like. My dad took me to see the sights. We climbed the Great Wall, holding silly kung fu poses that recalled the old Cantonese martial arts movies we used to watch together. We walked through Tiananmen Square. I took photos of the iconic portrait of Mao Zedong, and of armed guards who shouted in Mandarin as I approached, "Stop! Don't come any closer!"

In those pre-boom days in Beijing, my father lived at the end of a dirt lane, across the street from a shantytown where a pair of sisters washed their hair with a bucket of water dumped from overhead while chickens ran free at their feet. I snapped photos of this visual smorgasbord, eager for these glimpses of the world my father occupied without me.

That week, he proffered a hesitant invitation to paint with him in his studio, just like old times. We sketched, mixed oil palettes, sprayed fixative, opened windows to flush out the fumes. We whiled away the hours, painting the same old man's face.

Much later, he told me that my visit fell during a period in which he was struggling to make ends meet: *I was embarrassed. I didn't want you to see me that way.* This is what fathers want for their daughters.

As for me, I wish I could say that after that trip my father was back in my life fully: *Look at all the fun we had. It made us feel close again.* This is what daughters want from their fathers.

YOU WANT HIM to miss you, to want to see you more. You want him to make more of an effort to meet you halfway. And you want your brother to talk to your father too, so that you aren't always doing all the work.

But it's more like this: You go about living your life. Your father still doesn't express much interest in seeing you more than once every couple of years. You get married. Everyone agrees that it's probably better if he doesn't come to the wedding. You make the trip to see him, this time with your husband. It's a lovely visit, but it's a major effort—in time, in money, in emotional labor.

Eventually, you have your own children to think about. You bring your infant son to Hong Kong to meet your dad, and have your trip cut short by an earthquake and tsunami in Japan. When your second child arrives, you bring the both of them, traveling by yourself from a visit with other family in Japan, sweating through the diaper explosions and the interminable wailing, nearly missing your connecting flight, all because you *want him to know them.*

On that arduous trip, as a new mother of two, you think that his interest seems to go little beyond aesthetics. He finds the boys beautiful, enjoys watching them from a distance. But he doesn't really engage with them. You miss the version of him who loved to play.

When you return home, you give up—not all the way, not forever, but all the same it costs you something. The bitterness returns. Three years pass. But little things keep reminding you of him. Those Choose Your Own Adventure books your older son, now eight, is reading. Your father has met him only three times, the last time when the boy was five. That's the age your younger son is now. No one judges you. But then comes a movie suggestion on iTunes, for a long-ago film that features your father's painting—from a computer algorithm, of all things. Gently telling you what to notice about your own life, reminding you of your father and his art.

If you decide to go visit your father, turn the page.

SO MY BROTHER and I went to see our father—all three of us together in China, for the first time in nearly twenty years. We boxed in his studio, did a little sketching, made plans for him to come visit us in California. Andy said it was the first time he felt that our father saw him as a real person, and not as the child he was when our father last lived with us. We were happy.

And then, on the night before we were to leave, we watched as our father nearly died on the floor in front of us.

The new pull-up bar he'd been doing flips on had fallen off the wall. In the flurry to get the apartment ready for our visit, he hadn't securely bracketed the bar. This was an oversight that was hard to square with the past—we'd always had one of these bars, and our father had taught us to always take care and secure it. But it didn't really matter now.

My brother and I rushed to his side as our stepmother frantically called for an ambulance. Our father's eyes were open but unseeing; his breath was irregular, gasping; there was blood on the floor. Andy was a physical therapist now, his interest in the body and how it worked a direct result of the lottery draw of our dad being our dad. As he performed checks for consciousness, our eyes met over our father's body:

Is this why we're here? To be here when he dies?

When our father woke up, he didn't know us. He spoke to us in different languages. When the ambulance arrived, I called upon my twenty-years-expired first-aid lifeguard training and stabilized his head and neck injuries. Then I screamed at the paramedics in Cantonese that we could not shove my father into a wheelchair with the likelihood of skull fractures so great.

I answered my father's questions, repeated over and over again, on endless loop: "What happened?" "Where?" "I did?" At the horror-movie hospital, with its filthy floors, flickering fluorescent bulbs, mildewed pillows, and people with head wounds walking around while flicking cigarette butts, I gave the young ER doctor a piece of my mind.

But my dad didn't die. By the time my brother and I flew home, having moved him to a new hospital and leaving him in the care of our stepmother, he knew us again. Still, it wasn't yet clear how far back he'd come from where he was.

After the brain injury, as the months passed and we monitored his progress in the crisscross exchange of photos and messages across the Pacific, I knew he was going to be okay not when he started doing his exercises again, but when he started cracking jokes about it. *Am I fit or what?*

To have made it through the most terrifying moments of our

lives and come out with all of us intact—my father's health returned to him—seemed like a miracle. A gift not to be squandered.

The next year, I flew back to Guangzhou to see him again, the last visit before the pandemic. We drew and painted together in the studio, hiked in the hills, had long talks on the couch. I cried only once, when we had that old familiar fight about him coming to see us in California.

The day before I left, he brought out some photographs he'd found. Us painting together, at thirty-nine and eight, and at fifty-two and twenty-one. I knew then, without him having to actually tell me, that he was saying he loved me, no matter how infrequently we saw each other in person.

I've been drawing again. Keeping a sketchbook. Watercolors and pencil sketches of muscles and bodies; meditations about movement and being alive. It has been my way of staying on more than just nodding terms with him, with us, with the selves we used to be. With who we are now.

SKELETAL MUSCLE FIBERS are unique cells in the human body. They are long and skinny, run parallel to each other, and have multiple nuclei. We're born with a set number of those fibers. They grow bigger in response to exercise, not by dividing, but by recruiting muscle satellite cells—stem cells specific to muscle that are dormant until activated in response to injury—to contribute their own nuclei to muscle growth and regeneration.

Meaningful muscle growth, though, requires challenge and stress. It requires you to be active. If muscle cells aren't subject to harder work than they're used to, they won't grow. And they'll shrink and atrophy if you don't use them.

Every day, my muscles are reminding me of things I need to remember about being a person. Today the lesson is clear: Though we are essentially who we are, we're capable of change. Sometimes that process is painful, because, well, metamorphosis can be ugly. But we keep trying, so we can eventually get to something we find beautiful.

ACTION

Muscles are highly expressive; their individual action betrays a particular movement of the soul.

—G. B. DUCHENNE DE BOULOGNE

Is smiling a practice? And is joy a muscle?

—SARAH RUHL

8

Your Muscles Are Talking

One June day early in the pandemic, Dan O'Conor jumped into Lake Michigan.

That morning, the fifty-two-year-old Chicagoan was feeling especially beaten down by the worries of the world. He was also excruciatingly hung over, after celebrating his son's high school graduation the evening before, when a bourbon collection had been severely depleted. So palpable was his misery that his wife, Margaret, booted him out of the house for a bike ride down to the lake. When he got there, he stared at the water.

The act of launching himself into the air felt a little like lunacy, but it was also a momentary suspension from all his cares—and that, truth be told, felt terrific. So he came back the next day and did it again.

He kept jumping. Every time he thought he might stop, something would happen—like the time Dan hit day 150 and his friend tipped off a local news blog. Or when the AM talk radio guys invited him on the air and wondered aloud if he could jump through the Chicago winter. (He took that as a challenge.) Every once in a while, someone would come along and ask if they could jump with him. (They'd also ask if he was bananas, or suicidal.) There were days when he had to take a shovel and dig a hole in the snow and ice to jump into. And people began watching his daily jumps on social media, by the thousands.

He made a robe for himself, stenciled with GREAT LAKE JUMPER in colorful block letters.

A dedicated music fan, Dan had worked at *Spin* magazine for fifteen years; at Margaret's suggestion, he invited local bands to provide musical accompaniment for his jumps. He booked acts for what ended up becoming Musician Mondays.

On the one-year anniversary of the Great Jump, Jeff Tweedy of Wilco showed up to play, and sang Dan's favorite song by the band, "Spiders (Kidsmoke)." Tweedy even changed the lyrics to suit the occasion: "On a private beach in Michigan" became "On a public beach on Lake Michigan."

Something like a party took shape at Dan's favored takeoff spot, on the concrete lip of the Montrose Steps, with its idyllic view of the downtown Chicago skyline. After a solo leap off the top of a tall ladder—and over Tweedy's head as he sang—150 people joined in for a group jump. The crowd went through sixty pounds of pulled pork in two hours. At one point, Dan looked around and thought, *This is really joyful.* This time, he was not hung over.

He took day 366 off. Then he started all over again.

Most days, via social media, I watch Dan throw himself into the lake. I've watched countless clips of his jumps, many of them more than once, for no other reason than the fact that they mesmerize me into a good mood. The videos are brief, roughly twenty seconds. Most start with that view of the Chicago skyline and the edge of the lake. Then Dan will come flying into the frame, sometimes in slow motion: a boisterous leap, or a casual front flip. He doesn't have a huge repertoire; occasionally he'll do a backflip, a jackknife, or an inelegant belly flop. My favorite might be the sailor dive—a hilarious headfirst plunge into the water with arms held down fast alongside his body. The jumps always make me smile, and more than a few tickle enough to bring forth a loud guffaw.

When I get the chance to meet Dan in person in Chicago, he'd just passed the two-year mark on his jumping journey. While he'd managed to get out of town a few times in the previous year—road trips to jump into each and every one of the other Great Lakes, as well as leaps off "the Jaws Bridge" on Martha's Vineyard and a not-quite-complete flip off a twenty-two-foot-high houseboat that resulted in a bruised torso—he reckoned that he'd jumped into Lake Michigan about 750 of the last 780 days, give or take a few. The pandemic kept rolling, and he kept jumping.

On a late-summer morning after his daily jump, the glare off the water already baking the ladders along the concrete ledge off Lake Shore Drive, I ask him: "What's this all about?"

He laughs, the points of his mustache twitching with water droplets. "A lot of people have asked me over the last few years 'Are you still swimming?' And I'll say 'Well, it's not much swimming—I'm jumping.' And then other people will say 'How's the polar bear plunge going?' And then I'll say 'Well, it's not a polar bear plunge, because I do it in the summer too.'" He thinks for a minute, trying to settle on the right words.

"This is about me trying to fly," he tells me finally. "That's the thing I'm after. Even if it's brief—just one second—there's something magical about that."

JUMPING IS A basic test of physical fitness. In the NFL Scouting Combine, vertical leap is one of the measures used to determine who will ultimately be drafted to play pro football. No matter what the sport, it's a way to gauge power and athleticism—explosiveness off the ground. (Football scouts have referred to a prospective player's strong gluteus maximus by using the term "high butt factor.") Surfers, for example, might not be thought of as having exceptional vertical leaps, but the best of them have jumps as good as NBA point guards.

The medical director for the first-ever US Olympic surf team, Kevyn Dean, who worked on biomechanics research with Kobe Bryant, tells me that he loves the ability to measure athletes across different sports—say, surfers and basketball players versus ice-skaters and volleyball players.

"Did you know that the distance covered during a triple axel"—fifteen feet—"is the same as from foul line to basket in a slam dunk?" Kevyn asks. "Or at least that's what I've been told."

I picture a tiny teenager doing a three-rotation jump on skates in the same airspace as a six-foot-eight basketball player dunking on the court, and shake my head at the incongruity.

Kevyn grins. He has worked with all kinds of athletes on improving their biomechanics—in other words, he helps them learn how to jump better. After all these years, he is still awed by what the human body can do.

The best place for watching humans learn how to jump, Kevyn says, is a playground. "It's really fun to watch kids figuring out their bodies, working out how to jump and land, in real time," he tells me. "They often do it wrong. They might land straight-legged. You see them begin to understand that 'Oh, if I bend down, I can jump higher!' And then they try again, finding their feet in space, figuring it out."

Jumping is a very early movement in human development, as we learn how to interact with the world. It's extremely functional, but it's also just fun: *How high can I go?*

The human body has evolved to move: to lift, jump, climb, run. When our muscles need energy to move, the mitochondria in those muscle cells make it readily available in an instant—sending an electrical charge across membranes to convert nutrients into the energy-carrying molecule adenosine triphosphate, or ATP, which is then used to contract muscles. This is why mitochondria are called the powerhouses of the cell. Interestingly enough, the depletion of ATP, which is also necessary to release myosin from actin filaments to relax muscle, is what leads to rigor mortis after death.

Part of what makes movement feel good is the neurochemical signal cascade from brain to muscles and back again. Skeletal muscle allows you to move your body around; it's the largest organ in the human body, responsible for about 40 percent of your total weight. It is also an endocrine tissue, which means that it releases signaling molecules, which travel to other parts of your body to tell them to do things. The protein molecules that transmit messages from skeletal muscle to other tissues—including the brain—are called myokines.

Myokines are released into the bloodstream when your muscles contract, activate muscle stem cells for repair, or perform other metabolic activities. When they arrive at the brain, they regulate physiological and metabolic responses there too. As a result, myokines have the ability to affect cognition, mood, and emotional behavior—kind of like a love letter from your muscle to your brain. Exercise further stimulates what scientists call "brain-muscle cross talk," and these myokine messengers help determine what specific beneficial responses result in the brain, including the formation of new neurons and increased synaptic plasticity, both of which boost learning and memory. Bulking up muscles literally bulks up your brain! (This is why kids do better in school with PE.)

In essence, your muscles and brain are talking to each other all the time. They're sending good vibes back and forth, keeping each other healthy over your lifetime. Your body on exercise, even after just a few minutes, stimulates the brain, muscle, and other tissues to release a wash of molecules that change you: turning genes on and off, waking up immune cells, controlling inflammation, adjusting levels of blood sugar, ramping up metabolism, tissue healing, and more, like the release of endocannabinoids and dopamine, which help give you that overall feeling of well-being and euphoria.

The Stanford University health psychologist Kelly McGonigal has evocatively described the antidepressant molecules released by exercise as "a pharmacy in your muscles." If you don't move, lousy things happen: Research on bed rest reveals that, within days, the heart's output drops; arteries narrow and stiffen; and actin and myosin, the proteins that make up muscle, break down. Not only that, bed rest can be terrible for mental health—the agony that follows immobilization (see: recovering combat veterans; people with high-risk pregnancies) is well-documented.

Somewhere in the midst of this research on jumping, I find that I'm unable to jump, because I've injured my soleus, a muscle in the calf used for jumping, walking, and standing. It pulls against the force of gravity to keep the body upright. This is the result of an unfortunate chain of events involving my surfboard and my leg. The inability to get off the ground greatly sours my mood. But it does get me thinking about how pure movement is special, and how jumping is perhaps the most joyful movement there is.

RIGHT UP UNTIL he graduated from college, Dan O'Conor was the kid who did whatever sport was in season: baseball, basketball, track and field. He played football at the College of the Holy Cross, in Worcester, Massachusetts; during his tenure there, he tells me, the team had its winningest four years since 1891. Every summer, his family packed up and drove from Chicago to Cape Cod, where his parents were from. From dawn until dusk, he and his six siblings were in motion; more often than not, he whiled away the hours on a diving board or a swimming raft, practicing his jumps.

When Dan first jumped into Lake Michigan on that early pandemic morning, he realized that he'd missed taking flight. Even after thirty years, the flip came back quickly: *Duck your head, and let the momentum take you.*

Jumping into Lake Michigan is Dan's daily inoculation against inertia in midlife, but what strikes me most is how merely *watching* him jump has the effect of lifting the moods of countless others as well.

"Look, I'm a fifty-four-year-old dude in Chicago," Dan tells me. "I don't know if I'm doing social media correctly. This whole jumping thing is a succession of bad camera angles and bad lighting. But it's definitely something that brightens my mood. The positive feedback from strangers, and all these people reaching out—that surprised me. It's pretty special."

For me, too, observing Dan taking off in a flying eagle pose over

Lake Michigan is joy-inducing. When I answer him now in real time with a running leap of my own—a kind of movement-based call-and-response—it's a way to continue our conversation about how the body can lead in elevating the spirit. *I see you! And I will raise you one can opener!* (Another passerby, on a wintry day—observing Dan's preparations for cannonballing into perilous, wind-whipped thirty-five-degree chop—could be forgiven for worrying that he was doing himself ill.)

He is continually surprised by the connections it brings. A few days before we met, he arrived at his usual ladder to find a family of four from Kansas City hanging out. He was about to move down to the next ladder when they spotted him. "Are you the Great Lake Jumper?" they asked excitedly. "We've been following your jumps!" The father became the day's cameraman.

In recounting this story, Dan is a bit bashful, then thoughtful. "I'd never done anything three hundred sixty-five days in a row before, except maybe brush my teeth," he says. "I don't know how long I'll keep jumping, but for now, it still feels good to me."

We gaze out at the splendid skyscraper views: Here is the John Hancock Building, there is Lake Point Tower. Dan points out a relative newcomer, the interconnected wavelike pillars of the St. Regis Chicago, by the renowned local architect Jeanne Gang. "If I could jump over those tall buildings," he says, "I would."

9

Jumpology

It turns out that Charles Darwin had a thing or two to say about jumping. "Under a transport of Joy or of vivid Pleasure," he observed in *The Expression of the Emotions in Man and Animals*, published in 1872, "there is a strong tendency to various purposeless movements, and to the utterance of various sounds. We see this in our young children, in their loud laughter, clapping of hands, and jumping for joy; in the bounding and barking of a dog when going out to walk with his master; and in the frisking of a horse when turned out into an open field."

When we feel good inside, we are moved to move. Joy can be expressed through the body in muscular movements, but Darwin noted that the opposite is also true: The body itself can be a path to joy. Kids and puppies romp because, he wrote, "the mere exertion of the muscles after long rest or confinement is itself a pleasure." Joy lies in the stimulation of brain and body, and in the conversation between the two. More than 150 years ago, this was brain-muscle cross talk as Darwin saw it.

"The body is a powerful tool for communication," says Dacher Keltner, a professor of psychology who heads the Berkeley Social

Interaction Laboratory and is the founding director of the Greater Good Science Center at the University of California, Berkeley. His research focuses on the biological and evolutionary origins of emotion, and he's also a noted happiness scholar.

"There are certain emotions that are more about the body," he tells me. "Joy, for example, which often involves jumping. Or love, which is about embracing, postural movements. Emotions are about action. And the preparation and intention for action."

I'd never considered feelings as action-oriented, but once I do, I see it everywhere. The Dutch social psychologist Batja Mesquita has said that emotions can be thought of as "relational acts between people," rather than as simply mental states that reside within us. Emotions require *exertion*—they are expressed outwardly, so that they can make something happen. In addition, "many of our emotion terms are references to states of the body—we're downcast, bent out of shape, head over heels, shaken up, down in the mouth," according to the philosopher Nikhil Krishnan, who wrote a magazine story about Mesquita's work. Our muscles showcase our inner lives in more ways than we might realize.

Research on emotion has historically focused on facial expression rather than on other parts of the body. Are you smiling? Frowning? Clenching your jaw in a teeth-baring grimace? In fact, Darwin's own work on emotions was inspired by that of the pioneering French neurologist G. B. Duchenne de Boulogne, who used clinical photography and electrical stimulation to study the specific musculature of the face. He believed that muscles revealed the language of the passions, and his 1862 monograph was the first to study the physiology of emotion. If you're smiling, you're likely activating the zygomaticus major, which Duchenne dubbed "the muscle of joy." His wide-ranging studies, from the description of various muscle atrophies to the invention of an instrument for muscle biopsy, have had lasting influence on fields including neurology, muscular disease, plastic surgery, medical photography, and modern art.

More recent work on emotions indicates that they can be recognized from the body alone, even with very limited information. Studies show that jumping and expanding rhythmic movements in general often occur when we are happy. Observing momentary jumping movements on their own—without facial cues or context, and even without the actual intent of the mover to communicate a particular emotion—is enough for the observer to identify the emotion as happiness. There's enough evidence for scientists to believe that there is an underlying brain mechanism for emotion perception from body movements and actions that exists across cultures and studies.

In other words: When we watch Dan O'Conor jump into Lake Michigan, we recognize it as joy. The movement is the message. In the darkest days of the pandemic, this was a message he needed to send—and, it turns out, one many of us needed to receive.

In the Darwinian signaling world, jumping has powerful and dramatic meaning. "Darwin's descriptions of jumping are really cool—how we dance for joy, for example, and how the joyful parts of dance are really about jumping," Dacher explains to me. "And dance is emotion, culturally expressed."

It should come as no surprise that sports fall into this category as well, and Dacher points to his favorite game: basketball. "It's the quintessential example—there's a lot of jumping for fun in basketball," he says. "Kids do it, and adults too. Joy is when you leave your worldly concerns behind and feel free, and jumping is like that. You're off the ground and floating."

Here, my mind turns to baseball. Because there's a moment at the conclusion of pretty much every World Series championship that I find completely disarming, when grown men throw off all their mean mugging and stoical spitting in the dugout and race out to the baseball diamond for no other reason than to truly and ecstatically jump for joy. This team jack-in-the-box, as I see it, is the most notable play of the game. The guys turn to each other, jumping into each other's arms, jumping *on top of* each other, up and down and up and down, as if on springs, each unable to contain the excitement within, which *must come out*. It's communal jumping as celebration. Or, as Dacher prefers to call it, "collective effervescence."

LET'S TALK ABOUT whales for a minute. When it comes to phenomenal displays of fitness, one of the largest living animals on Earth propelling its fifty-ton body out of the water is hard to top. Whales often move en masse, jumping together, almost as a single organism. Scientists who study the physics of breaching have described the whale jump as "an iconic animal behavior that tests muscle fibers' physiological limits." The extraordinary energy required for this movement likely represents the single most powerful burst maneuver found in nature.

A whale breach requires swimming at high speed in the lead-up to the jump, and the contraction of a muscle called the peduncle, which connects the tail flukes—the two halves of the tail—to the dorsal, or backside, fin. You can think of the peduncle as the whale equivalent of our gluteus maximus: a butt muscle used for jumping. It may well be the strongest muscle in the animal kingdom.

I've had the profound pleasure of observing humpback whales breaching from the perspective of a boat (and from a surfboard!), but it's hard to know exactly what happens below water level. To get a sense of the underwater effort leading up to the jump, I track down dorsal camera-tag footage of a fifty-ton humpback ascending to the surface to breach—to be most clear, that's *one hundred thousand pounds* of whale, or the weight of a large Gulfstream jet.

If you've ever wondered what it's like to ride on the back of a whale, this view comes pretty damn close: First, the whale's head undulates to begin the climb. After several full-body up-and-down strokes, the camera begins to vibrate as the whale picks up speed. The green water clears and the sun's orb becomes visible as the whale bursts out of the sea and into blue sky. The whale exhales a plume from its blowhole, right before splashdown. The whole thing takes about twenty seconds, and it's exhilarating.

A fifty-ton whale breaching just once expends energy roughly equivalent to what I'd need to run an entire marathon, and whales breach often. So why do they do it? That's the subject of various theories; research suggests that it's about communicating location or activity or to signal fitness—and to play. Through repeated breaching, humpback whale calves are also thought to be boosting myoglobin, the protein that binds iron and oxygen in the muscles, to improve diving ability.

Because breaching uses up so much energy, the fact that mothers and calves often breach repeatedly side by side, far from feeding grounds, indicates that it must serve a special and significant purpose for young whales. Not that you can get inside the mind of a whale, but almost every research paper I have read about breaching notes that juvenile whales seem to *really like it*. I'm reminded of Darwin again: dogs bounding, horses frisking, whales breaching, humans jumping, all in the pursuit of pleasure.

In this buoyant spirit, I turn to a favorite passage in Annie Dillard's classic *Pilgrim at Tinker Creek*, in which she described walking around a pond, the whole of it "popping with life." One frog, Dillard observed, big and viridian-bright "like a poster-paint frog," declined to jump, so, she wrote, "I waved my arm and stamped to scare it, and it jumped suddenly, and I jumped, and then everything in the pond jumped, and I laughed and laughed."

MY FATHER ALWAYS jumped rope with a measured economy. Through the door that led from the house to the garage, I could hear the steady *whupuwhupuwhupuwhupuwhup* of the leather cord. It spun so fast that it wasn't even visible as a cord, just as a blurry force field that he conjured up around him with the wooden grips in his hands.

Part of the magic was that he never seemed to put forth much effort—feet light in their left-right-left-right stepping, breath calm and efficient, face still, even as he'd throw in a double: two turns of the rope under a single jump. Then, and only then, would his face crack, to reveal that familiar big smile.

My father didn't jump in community. For him, jumping was a solitary practice, more like meditation. He'd jump for an hour or so in the garage, the whip of the rope hitting the concrete floor like a metronome. The drip of his sweat was like a watery echo, eventually collecting into a puddle at his feet.

I jumped with a red-and-white-striped cloth rope that didn't spin as fast as his, but it didn't sting as much when it hit my bare legs either. Though I worked hard to emulate my dad's light steps, mostly it felt like my feet were plodding in heavy time: up, then *down*, up, then *down*. Gravity kept intervening, with annoying insistence. Learning to jump one-footed was a meaningful achievement. It was moving to a faster beat, a new cadence of keeping time, to hover like a hummingbird.

These time changes call to me now. So I start jumping again, humbly, five minutes at a time, 750 jumps, as many days as I can remember to do it. I change up the pace, running my feet, trying something new. I play. When I jump, I listen to the whir of the rope. Sometimes I close my eyes. I count until the count seems to go on without me, looping back to zero every time I reach another hundred. The number doesn't matter so much as what that loop holds inside it: a rhythm I control; the pleasant *tickatickaticka* of the rope on the wooden floorboards; me, in suspended animation, for as long as I care to be there.

I LIKE DARWIN'S idea of using the body as a tool to find levity. Movement born of muscle is its own language, intrinsic to the body. How we use that language to communicate with others has its own power. I seek out other jumpers, because I want to feel, in their presence, how jumping is essential to their lives.

When the dancer and choreographer Ebony Ingram was growing up on the South Side of Chicago, her older sister and cousin taught her and her younger sister how to jump double Dutch. They'd spend hours outside, taking turns jumping and turning with a rubberized clothesline, making up songs and rhymes. "As Black girls, that's a part of what we do—it's how we commune with each other," she explains to me. Strangers would come and ask the girls to teach them or to let them try a turn. More than anything else, double Dutch was about connecting with people.

A little over a decade ago, Ebony moved to Washington, DC, to attend Howard University. She remembers a day in Malcolm X Park, when she was new to the city: "My sister and my friends and

I, we decided to go jump in the park, and people were jumping out of their cars to play with us." Her choice of "play" to describe the act of jumping double Dutch is no coincidence.

"There's something about it that's ecstatic," Ebony says, her face animated, her hands flitting expressively around her close-cropped curly hair. "You feel it in your chest, the excitement and joy, succeeding in it." At heart, double Dutch is a game, but musicality and style elevate it to an art form, with a choreography all its own. And the complicated jumps, spins, steps, and acrobatics executed within two swirling ropes, keeping time all the while, requires sheer athleticism. The fastest double Dutch jumpers average almost two hundred jumps a minute.

A few years ago, Ebony was the double Dutch coach for Joshua Wilder's play *She A Gem*, a world premiere commission by the John F. Kennedy Center for the Performing Arts, about four teenage girls preparing for a double Dutch competition. Double Dutch became a competitive sport in the 1970s; these days, there are many professional teams that teach it. Every year since 1992, an international double Dutch competition has been held at the Apollo Theater in New York City, attracting teams from France to Japan. Despite the fact that double Dutch has gone global, it has its roots in neighborhoods from Harlem and North Philadelphia to Chicago's South Side. It's where the sport developed into what it is today.

"With the professional teams, there's a standard, a certain way of doing things," Ebony says. "What I do is different because it's street jumping. That's what double Dutch is. And the rope is a character. The rope is alive."

The ethnomusicologist Kyra Gaunt has described double Dutch as an oral and kinetic tradition. Learning the language of double

Dutch—getting in the rope, learning the tricks, feeling your body, relying on intuition—is a practice in confidence. "The rope knows if you hesitate—it'll smack you in the face," Ebony says. "It's a good teacher, in life, to just go for it."

The community aspect is critical. There are rules and an order: Turners adjust the rhythm and pace of their twirling to the jumpers inside the ropes. There can be a hierarchy; if you're younger than the rest, you may be turning the ropes for everyone else before you get to jump.

"But if someone skips you, the energy changes in the game—people start looking at each other," Ebony says. "It's an exercise in getting along."

SO IT IS with Ebony in Washington, DC, that I receive my initiation into the rope. The trees huff off their leaves in bright-colored breaths, and the late-afternoon light casts a warm glow. At Martha's Table, a modern glass-walled community center with gardens at the top of a hill in the southeast part of the city, Coach Robbin Ebb and her sister Carlyle Prince, a.k.a. Cee Cee, are coaxing me in. In their expert hands, a pair of blue-and-white beaded ropes beat a steady *tickticktickticktick* rhythm. I listen and watch, bouncing on my toes.

Ebony cheers loudly from the bleachers; we came here today so I could learn from the best. Joy Jones, the ebullient sixty-eight-year-old founder of DC Retro Jumpers, laughs and takes a video of the scene with her phone. Over the last two decades, Joy's program has taught thousands of people to jump double Dutch, at schools, community events, and festivals around the greater DC area.

I like jumping rope, but I've never jumped double Dutch. The swirl of a double rope is daunting, as is the crowd of people watching and the ever-increasing queue of those waiting for a turn. Here I am, a grown woman, made to feel like a kid again. The moment of entry is the most intimidating of all: me, a still point in this turning-rope world, readying myself to leap into the blur.

Coach Robbin has chalked a heart onto the floor, so that beginners will know where to jump in. She gives me my cue: "Bunny hop, bunny hop, bunny hop, go!"

I take a deep breath, jump forward under her turning arm, and aim for the heart.

"Apple on a stick, make me sick, make my heart go two four six . . ." Cee Cee's rhymes make me laugh, but they also keep time and give me something to focus on.

Behind me, Coach Robbin offers suggestions for different footwork: "Bunny hop! Crisscross those toes! Turn around and skip! Turn around and skip! Run, run, run, run, run!"

The rhymes quicken; so do my heart and feet. As I jump, the smile on my face grows wider and my green skirt flies higher. Spectators shout encouragement, the sound surging louder as the rope picks up speed. Finally, when the rope hits my feet, the crowd claps and hoots in appreciation.

This is joy, on repeat—for tiny, pigtailed Ella, who comes every week with her mother and grandmother; for Jalen, an elementary-school-age boy who wows the crowd by dropping and doing push-ups, all from between the ropes; for a group of teenagers in an after-school program, who surprise me with their shyness but dare each other to give it a go. Even for the parents and caregivers themselves who are lured into the electric thrill of the rope. Nobody is immune, and that's the point.

Joy's childhood took place largely in DC's Michigan Park,

a middle-class Black community where, she tells me, "us girls jumped rope day in and day out." But in 2004, when she started the DC Retro Jumpers as a team of eight middle-aged adults with weekly jump rope sessions, she hadn't jumped in decades. The pleasure and elation of rediscovering it as an adult was, she says, "like champagne and confetti."

"You get high, and you want to celebrate at the same time," she says. "I never liked going to the gym, which feels more like work. Something like double Dutch is more like playing."

We watch Coach Robbin and Cee Cee patiently work through the line. "Robbin likes working with children, but I like working with adults better," Joy says as we observe three-year-old Ella excitedly coaxing her own mother into the ropes for the first time. "Adults aren't used to playing anymore. I like seeing the transition—they're watching us from a distance, I spot them with that look on their face, they're inching closer. But they're embarrassed, even as they come up to the line. They say, 'I'm too old to do this. It's been too long.' It's their defense mechanism. But then they look at us"—here, Joy cracks up; she has the most marvelous full-body laugh—"and they know that excuse is not gonna work! They need coaxing and encouraging, especially if they're nervous about it. Then, at the moment they're jumping, they're so excited. The physical exercise, watching them light up inside with endorphins, them realizing they're doing something new, the joy on their face?"

She nods and points a finger skyward. "That's what turns me on."

Play is where you have the essential pleasure of feeling at home in your body. And why should you give it up just because you're a grown-up? As the afternoon deepens into evening, it becomes clear to me that the act of asking someone into the ropes is really an invitation to joy. Ingrained in the culture of double Dutch—*bunny hop, bunny hop, go*—is a generosity of spirit.

"TO MOVE THINGS is all that mankind can do, and for this task the sole executant is a muscle, whether it be whispering a syllable or felling a forest," said the biologist Charles Sherrington, during a 1924 lecture at Oxford University. The motor system controls more than six hundred muscles, and coordinating these muscles is carried out mostly without conscious instruction. Some actions are innate, but others have to be learned through practice. Most of what the motor system does is invisible to us until something goes wrong, through injury or illness.

Though Ebony had invited me into her beloved local double Dutch scene, it had been some time since she herself had jumped. In 2021, she was diagnosed with leukemia, and she received a stem cell transplant that August. Playing double Dutch as a kid had built the body she has as an adult. But after being in the hospital for months at a time, she'd had to walk at a slower pace; she still couldn't quite run. It had been hard for her to see what it would take to get from where she was to a body that jumped again.

Today, though, is the day. "I feel my muscles again," she says.

Coach Robbin and Cee Cee start the blue-and-white ropes ticking. Ebony stands to the right of Robbin's shoulder, her striped hat jaunty on her head, her whole body already dancing—it knows what to do. She asks them to turn a little faster. As she snaps her right arm in time and gets ready, I see how alive and alert she is. And when she finally jumps in, it is full tilt, with fury and speed, both legs pumping.

When you come into community with death, Ebony says,

"there's an urgency to get back into the rope, into the work." After chemo and radiation, jumping is not the easiest thing anymore. But she has every intention of being better than she's ever been. "I still need to jump," she says.

I still need to jump. This stays with me.

A body in motion doesn't always stay in motion, especially after it's knocked down by illness or by accident. But if you're lucky, jumping can be a guiding daily rhythm by which to mark time. It's a reminder of what it is to be on Earth now—and what it means to leave it.

THE LEGENDARY PHOTOGRAPHER Philippe Halsman had a record 101 *Life* magazine covers to his name. Over the long arc of his portrait-making career, he discovered that getting people to show their true selves boiled down to something very simple: asking them to jump.

"In a jump, the subject, in a sudden burst of energy, overcomes gravity," Halsman wrote. "He cannot simultaneously control his expressions, his facial and his limb muscles." In this concentrated, whole-body movement, he said, all artifice falls away, and the "real self becomes visible."

In 1952, after a disappointing photo shoot with the Ford family to celebrate the automobile company's upcoming fiftieth anniversary, his last-minute inspiration to shoot family matriarch Eleanor Clay Ford jumping saved the day. (She took off her high heels.) He began regularly including jumps in his photographic practice thereafter.

Halsman's love of jumping dated back to his boyhood days in Latvia. He was an excellent jumper, skilled at backflips on the beach as well as easy, effortless leaps alongside his subjects to make them more comfortable with his unusual request. He didn't just ask athletes to jump—he asked all kinds of famous people to jump, from Richard Nixon, John Steinbeck, and the Duke and Duchess of Windsor to Marilyn Monroe, Lena Horne, and Salvador Dalí. His extraordinary image of Dalí took twenty-six takes, each one requiring the simultaneous tossing of three cats and a bucket of water.

Nearly two hundred photographs of famous people jumping are collected in *Philippe Halsman's Jump Book*, first published in 1959. The fact that his subjects all said yes reveals something essential about the motion: It is an act of pleasure.

"I realized that deep underneath people wanted to jump and considered jumping fun," he wrote. He saw the body in midleap as perhaps the most telling of a subject's character. He even had a tongue-in-cheek name for his personal Rorschach test: jumpology.

"Jumping humanity can be divided into two categories: one which tries to jump as high as possible and one which doesn't care," he wrote.

In a playful nod to Freud, Halsman made further observations in a section titled "The Interpretation of Jumps." A person who doesn't use hands and arms during a jump? This is a person who doesn't like to communicate—or is unable to. A person who points with an outstretched arm? A sign of ambition. A person who jumps with arms out to steady them? This is a person who feels insecure or fearful in life. A man who jumps with legs apart? Here's a guy who tries to project strength and importance. A woman who jumps with legs apart? A sign of passion, independence, and rebellion.

His oldest jumper was the venerable Judge Learned Hand, who, at eighty-seven, murmured thoughtfully as he prepared to jump, "After all, maybe, this is not a bad way to go."

Later, Halsman would look at the picture of the jumping judge and reflect upon his own aging self with markedly less trepidation and a great deal more inspiration, writing: "I hope it will mean to the judge what it means to me—a proof that the secret of eternal youth is in one's spirit."

There's a digital picture frame on the desk in our family room, where I like to write. Over several weeks of writing, daydreaming, and staring off into space—and directly into that frame, with its rotating library of pictures—I realize that mine is a jumping family. Photos of my children catapulting off playground structures and diving boards. Slow-motion movies of them trying to get off the ground, brows furrowed in concentration, their plump cheeks bouncing as they land. My husband and me, high above a trampoline at a friend's birthday party, laughing uncontrollably. Various of us leaping to catch a Frisbee, soaring off a starting block at the pool, getting air off a surfboard. Like Halsman with his photos, I see personality in those leaps, an unfiltered mix of caution and glee. They are our un-self-conscious selves revealed.

A friend tells me that when she was a child, she found jumping on a trampoline relaxing. The rhythm calmed her hyperactive mind and organized her jangled body. A geriatrician tells me that the impact of gravity-related activities is critical for maintaining bone density, muscle strength, and balance as we get older. Just as there are biological underpinnings to whales breaching, jumping, it seems, includes ingredients essential to the longevity of our own human bodies in motion.

Halsman's theory of jumpology is part psychology, part

biomechanics—character analysis writ by muscle. I'm taken with the idea that our character is somehow manifest in our movements. That we know each other from a whole gestalt of small tilts, shifts, and gestures that make up the shape of us. I begin to think of this as a kind of motion gestalt: a pattern of movement that is all our own.

10

The Movement Is the Message

Many times in recent years, I have jerked awake, stuck in a recurring nightmare that has me frozen, pinned in place, unable to move *and about to die*. I see it coming, and yet I can't stop it—I am suddenly helpless, unable to exert my will on the world. I don't put a lot of stock in dream interpretation, but this one doesn't need much parsing.

In quiet desperation, I find any way I can to move, to find joy in movement, to feel situated in my days. Exercise, then, is a way to grapple with existence, to act. The daily practice of movement is something that helps us to develop presence.

I didn't take up surfing until I was nearly thirty. But there's a Polaroid my mom took of me as a pigtailed four-year-old surfing my dad's knees: arms out, athletic stance, a complicated look of anxious elation working its way across my face.

The image is slightly blurred, but I can make out the yellow polka-dot print on my pajamas. I suppose it's more evidence of the kind of dawn-till-dusk active existence my father wanted to shape for us—a life spent in perpetual motion.

Left foot, right foot. Left over right, forward, then back.

Surfing has become my daily practice, and as I work out timing, coordination, and proprioception—the sense of the body in time and space—I come to understand it as a kind of dance. I think about my brain in conversation with my muscles, telling them what to do. The concreteness of it comforts.

Sometimes my body moves in the opposite direction of my surfboard, and I go somersaulting, spread-eagled in the most awkward of ways: *How did it go wrong?* And sometimes it all clicks into place, and the board is an extension of my body, and I'm flying down the line without conscious effort, turning as wave and water dictate. That's when I think about nothing at all, except for the feeling of flow, happy for my dance partner.

In whatever physical practice you choose, there are moments when you enter what the performance coach Brad Stulberg calls a "brave new world."

Movement, Stulberg has written, "demands you pay close attention to the signals your body is sending. Do I speed up or slow down? Is this merely the pain of arduous exertion, or is this the pain of a looming injury?" The specificity of the feedback—the water on my fingers, the burn in my shoulders and back, the board under my feet—helps to refine that practice.

"Keep doing this," he added, "and your ability to pay close attention—not just as it relates to your body, but to all of life—improves."

WHEN I DRIVE to the surf in the predawn darkness, I often think of my father. Already he is a decade past the age of his father when his father died. But the close calls he's had make me fear the time. I feel the clock ticking away, ahead of me, on the other side of the world. We never know when the body will fail us, no matter how strong we try to make it.

It has always been difficult to reconcile my father's absence from my adult life with his constant presence in my childhood—he was always home, always *there*, ready to spar, to play, to draw. Once, when I was twelve—during a perfectly average afternoon spent hanging out together, in the middle of a lengthy, meandering conversation conducted from different corners of the living-room sectional—he turned to me and told me I was his best friend.

The truth is that my father is most himself when we are in front of each other, physically present. He's not a talker or a writer, but a doer. He hates video chats; he's not great on email or text either. The asynchronous nature of written communication doesn't suit him. But in person, the long pauses in his speech are filled in with a comic flick of his eyebrow, a squeeze of my arm or hand, that big, silly laugh.

I've begun calling him on these early-hour drives to the surf, when it's evening where he is in China. In the darkness, if I listen to his voice on the car speakers and keep my eyes on the road, I can almost pretend that we are riding in the car together. Sometimes I get mad at him, for living so far away, for not trying harder to stay in touch.

It's an old story, because I don't think I ever got over him leaving us. He showed me what it was to jump and move for joy, but when he went away from us, I couldn't see him anymore. He'd jumped too far. I've spent the last two decades trying to figure out how to put the joy back in—how to have a relationship with someone who has become more idea than real flesh-and-blood person.

The other day, my son called me over to look at a photograph:

My father and me, boxing, him holding up a pair of pads as I throw a right jab. I'm drenched in sweat, and my glove has just made contact.

As I examined the image, taken on the last visit I made to China, it struck me again how my father has always communicated with us best in the language of the body: a good-natured sparring match, a life-drawing session, a breath-stealing hug.

The body speaks the language of action. There are no words to adequately describe the movements, because the movements *are* the language. They are the communication between one body and another. The physicality is the point. You and me, in this space, together. The movement creates the memory. It is a way to be together.

I've thought about this a lot these days, as my dad is perpetually that half-turn of the globe away from me. But the quotidian physical movements of my life—stretching, skipping rope, even the simple push-up—are foundational. They connect me to my father across time as well as space. Something is encoded in these daily practices. Maybe it's that every movement is a kind of palimpsest: holding the memory of movements past.

Any dancer will tell you that movement is a way to tell stories. The New York City Ballet dancer Russell Janzen, upon retirement, wrote about how his specific physical capabilities—"the particular functioning of my muscles, joints, bones, and tendons"—have been essential not only to his livelihood, but to his happiness. "In moments of alignment and control," he wrote, "my body is not just the vessel for my self-expression, it is the expression itself." He didn't just dance; he *was* the dancing.

To get between the ropes in double Dutch, say, is a communal act, a multigenerational conversation, with the rope and the rhyme creating their own music—it's a lesson in how movement is its own language, culture, and memory. You simply have to see those bodies moving in space to understand what is being expressed.

I look for people and places to show me what I already know: *The movement itself is the message.*

In movement, I remember I have a dad. The love of movement I have is really about my love for him. It's the language he taught me. We keep practicing on our own until we can speak it to each other again.

WHAT I HAVEN'T yet told you: As I am writing this, my husband's own father is losing the ability to use his muscles.

Three years ago, he was diagnosed with amyotrophic lateral sclerosis, or ALS. *Amyotrophic*: literally, "without muscle feeding" in Greek.

The weakness started in his hands and feet. It took some time for doctors to figure out why. The disease destroys nerve cells in the brain and spinal cord that control the body's muscles, eventually affecting the muscles needed to move, speak, eat, and breathe. When those motor neurons die, they stop sending messages to the muscles, which can no longer function. There is no cure.

My husband crisscrosses the country to see his dad as much as he can, but our children haven't seen their grandfather in close to a year. In that time, he has lost the ability to walk. From the perspective of the phone camera, he converses normally, cracking brief jokes and asking the boys how school is going. Teddy and Felix, who are nine and eleven as of this writing, know the nature of their grandfather's illness. Still, on the phone, it's easy to pretend that nothing is different. After we hang up, though, I see Teddy's eyebrows knit together with worry, his corrugator supercilii—what Duchenne de Boulogne called the "muscle of pain"—telling me its own story.

"Grandpa looks different, doesn't he?" I say, and watch the space between his eyebrows relax again with this acknowledgment.

He sighs, then nods. His breath is a release.

All of us, at some point, are faced with limitations to movement. I begin to understand Galen's philosophy even more as one that applies to us, *now*. The feeling of not being able to move as we are accustomed, of our own volition, leads us to an existential question: Who are we then?

One afternoon not long afterward, Felix wanders into the room where I'm reading. "I just learned about locked-in syndrome," he says. "Basically, it's where your mind still works but you can't move at all except to blink, and it sounds terrible."

His is a rational mind, very much preoccupied with the scientific and technical elements of life, less so with their emotional valence. And yet there's a note of bewilderment in his voice.

"What would be the hardest part?" I ask gently.

"Waking up?" he answers. "All we do in life is move."

FLEXIBILITY

He has changed externally but he has not vanished.

—CARLOS MONTEZUMA

11

Muscles, Fast and Slow

James Francis "Jumbo" Elliott is considered one of the greatest track-and-field coaches in history. He was the track coach at Villanova University for more than three decades; he'd get up early and put in a day's worth of work at his heavy-equipment company before heading over to campus to lead practice in the afternoon. He sent twenty-eight athletes to the Olympics, and five of them came back gold medalists; his teams won eight national collegiate team titles; his athletes won eighty-two individual NCAA titles and set sixty-six world records.

When Jumbo himself was an undergraduate at Villanova, he was undefeated in the quarter mile in dual competition and one of the best in the world at that distance. He had hoped to compete at the 1936 Olympics in Berlin, but a pulled hamstring—one of the three muscles at the back of the leg vulnerable to hyperextension and tearing—got in the way.

Jumbo Elliott, who died in 1981, was also my husband's paternal grandfather.

Recently, Matt asked his dad, Jim, if he had any good muscle stories from Jumbo—sports, of course, being the love language of many fathers, and this being especially true in their family, and in mine. Jim's immediate answer: Frank Budd.

Frank Budd ran for Jumbo, arriving at Villanova in 1958 with what Jumbo referred to as a "good leg and a bad leg"—he had a weak right calf, likely because of polio. (When Budd was a boy, his mother applied a concoction of goose grease, nutmeg, mutton fat, and witch hazel to the leg, in an attempt to cure her limping child.) His left calf was significantly larger in circumference than the right, and Jumbo saw that when Budd ran, he still limped. "I guess I just didn't notice," Budd would later tell a reporter, "but it gradually grew stronger."

As Jumbo and Budd worked to equalize the muscles in his legs, Budd got faster and faster. In 1961, Budd ran 9.2 seconds to break the world record in the 100-yard dash at the national AAU Track and Field Championships at Randall's Island, New York. He accomplished this feat while running on the chewed-up inside lane of a cinder track, and felled the previous record of 9.3 seconds that had stood since 1948.

For the next two years, Frank Budd was the world's fastest human. In 1962, he was drafted by the Philadelphia Eagles, despite the fact that he hadn't played football since his high school days in New Jersey.

When I look further into the story of Frank Budd and Jumbo, I discover an open window into the Elliott house in 1962: a remarkably detailed "day in the life" profile of Jumbo in the *Sports Illustrated* archives. As I read the magazine article, I experience a kind of time travel: There I am at the breakfast table in Haverford, Pennsylvania, listening to my father-in-law as a fourteen-year-old high school freshman, talking to his father about Frank Budd.

"Dad," Jim says, "do you think Frank Budd will turn professional?" (This in response to the exciting news that Budd had been drafted by the Eagles.)

"Why, that's up to Frank, Jimmy," answers Jumbo as he sips his coffee. "I think he could make it in pro football if he wanted to. Of course, I'd like to see him stay eligible for the next Olympics. I'd like to see him stay on at Villanova and study law. I think Frank would make a very good lawyer. He could turn pro and still study law. He's thinking it over."

When asked by the reporter if he thought Budd could equal his own world record in the 100 yards, Jumbo's caffeine-fueled thoughts are clear.

"Can he equal it?" Jumbo says. "I think he'll break it. He ran that 9.2 100-yard dash under what I consider a handicap. He drew the first lane, which isn't as fast as the third, fourth, or fifth lanes on any track. Frank himself is much stronger than he was. He ran more last fall, he can take more work than he could in his first three years at Villanova. He has been doing a lot with the weights, working particularly to strengthen that right leg of his. Frank had polio as a child, and his right leg is about two inches smaller at the calf. But with this additional work, he is helping his right leg and smoothing out his technique generally."

Before he finishes his coffee and heads out for the day, Jumbo offers one more thought: "Frank is one of the fastest starters in the history of track, and with a little more endurance I think he'll be one of the real great, great sprinters of all time."

Frank Budd went on to set another world record, in the

220-yard dash, before hanging up his track shoes to play football. He spent two years in the NFL, in Philadelphia and Washington, DC, then played three more years in the Canadian Football League.

In 1996, Budd was diagnosed with multiple sclerosis, an autoimmune disease that attacks the protective myelin sheath of the nerves, often resulting in muscle weakness in the arms and legs, and, at the time, eventual paralysis. But he accompanied both of his daughters down the aisle, and put aside his walker to dance with one at her wedding in 2003.

"His life began with polio, the paralytic curse of three generations," declared a sportswriter for the *Star-Ledger*, in the New Jersey paper's late-life appreciation of Budd. His life would end with multiple sclerosis, but I think it's accurate to say that Budd used his muscles fully to express himself throughout. The fact that he became the fastest man on the planet was, as the sportswriter put it, "not so much an irony to Frank Budd as it was a lesson of what an iron will can do for a man."

IF MUSCLES ARE the organs of the will, what happens when that connection is lost or broken? In order to really explore how muscles work, I need to find out how they *don't* work. We inhabit different versions of our bodies over our lifetime. I think a lot about this now, as I've watched my father recover from a traumatic brain injury, my father-in-law confront ALS, and my mother struggle with a rare autoimmune condition—exacerbated by, of all things, the state of *not having enough muscle*. Muscle is a resource we draw on for more than we know. But in the face of change and loss, I admit to feeling confusion, and a measure of existential despair. How can paying attention to muscle solve for the unsolvable circumstances of injury and illness?

It's while in this frame of mind that I meet a yogi named Matthew Sanford.

Back in the fall of 1978, Matthew Sanford was thirteen years old and a gifted athlete, perpetually in motion. He loved all sports—baseball, hockey, golf—but basketball was the one he'd recently dedicated himself to full throttle, playing as he did on not one but three different teams. His memories of his last basketball season were the stuff of middle school hoop dreams. Being in the dribble, sprinting up the left side of the court, pulling up suddenly, the feeling of certainty as the ball left his hands: He already knew it was going in. He was a lanky, crafty point guard, full of potential, and happiest when reading the action on the court.

The head-to-toe hum of a jump shot: It starts with compression into the feet, a gathering of energy. And then an uncoiling, a full-body extension to release that energy through the fingertips to send the ball away. It's a rising up of strength that connects through space from body to hoop. In the fullness of that feeling, as young as he was, he knew what grace was.

Late that November, on a cold, misty Thanksgiving weekend, Matthew was asleep on the car ride home to Duluth, Minnesota, from the holiday spent at his aunt's house. The car, also carrying his parents and siblings, hit a patch of ice and tumbled down an embankment. His father and his sister were killed.

When he woke up in the hospital, he found himself fighting the respirator for control of his own breath. The feeling of being a "body ghost"—disconnected from his body in the most profound way—is a signal memory that he has returned to again and again in the years since. The catastrophic accident left him paralyzed from the chest down.

"For a long time, I missed being in my whole body," Matthew

tells me. "Because I was instructed to make the muscles of my upper body strong—to overcome my disability and drag my paralyzed body through life."

He tried hard to be a good patient, to forget the young athlete he was. During his recovery, he was told over and over that there would be no sensation below his point of injury—and if there was, it was just in his head.

This proved not to be entirely true. Though he didn't have direct control of the muscles below his chest or feel surface sensations on his skin, he could feel what he came to think of as a kind of tingle or hum: subtle sensations that gave him an indication of how his body was doing. Sometimes, when he moved in a certain way, he felt a buzz that ran all the way down to his toes.

It reminded him—of all things—of basketball.

He stopped mentioning it to his doctors. A decade after his injury, when he was twenty-five, he began to practice yoga.

CONSIDER THE WORD yoga. In Sanskrit, it means "to yoke," to dissolve separation between body, mind, and breath. In its ideal form, the practice is all about connection, and about being more aware of your body—to know it better and to recognize the parts you routinely ignore.

Matthew Sanford is a pioneer in adapting yoga for people with disabilities, like spinal cord and brain injuries, multiple sclerosis, ALS, muscular dystrophy, and cerebral palsy. A renowned teacher to students without disabilities as well, he is trained and certified in the Iyengar method, a style of yoga that focuses on precision and alignment of the physical body through the practice of individual asana, or poses.

Like many of his students, Matthew uses a wheelchair. But yoga taught him to resist the convention in the medical world to forget the parts of his body that were paralyzed; instead, he sought synthesis. He describes the practice of yoga as putting muscular action in service to the whole.

Asana, he says, helps restore shape to every body. When you put yourself into a pose, you gain strength, flexibility, and balance. You breathe better, and your entire body feels more at ease. The popular Western conception of yoga involves a specific kind of flexibility: a Gumby-like body that can stretch and contort into otherworldly postures. But yoga, an ancient discipline tracing back thousands of years in India, is fundamentally about grounding the body in this world, now.

Periodic disconnection is something each and every one of us experiences, Matthew says—whether we're paralyzed or not, and often on a daily basis. "Think of the contrast between slouching in your chair—with your sit bones like butter, your legs and lower back are dulled—and sitting up straight, at the edge of your chair, with your sit bones like knives," he explains, by way of example.

Instinctively, I find myself straightening up and scooting my butt to the edge of my seat.

"When your feet are on the floor and your head is stacked above your spine, your legs wake up—they are more alert," he says. "So are mine!" I watch as he repositions his legs with his hands. "There is connectivity to be found in alignment and precision, and in grounding the body—and that's especially important to the disabled body."

Yoga requires you to sit in the presence of the body you have. To feel more, and to feel more whole. In its practice, you are choosing—every time—to begin anew, to reestablish your body in the world.

BEFORE I ACTUALLY met Matthew in real life, I met him online, in his regular Monday morning yoga class. It was a vividly green spring day in Minneapolis, and he opened with an enigmatic lyric from a Talking Heads song:

Lost my shape, trying to act casual

Matthew, a tousle-haired man in his fifties with a boyish smile and friendly blue-green eyes, leaned toward us from a seated pose and grinned.

What does a Talking Heads song have to do with yoga? Doing all that is required of us—sitting for long periods, or even just getting a good breath—is fatiguing, he explained, and this all contributes to a loss of shape. I perked up at his matter-of-fact intention for the class: "We're going to have you exert your muscles a bit, to regain your shape."

I admit to being an intermittent practitioner of yoga. In my twenties, as an impatient young person living in New York, I sped through vinyasa flow classes to get strong and to sweat; in my thirties, when I was pregnant with my first child in San Francisco, I consented to the gentle pace of prenatal yoga as a way to find relief from the changes that confounded my body. Now in my forties, as an aging athlete who wants to feel good in my body for as long as possible—the litany of ailments growing ever longer, from broken ribs and stress injuries to chronic tightness or recurring pangs from old tears or sprains—I have become more acutely aware of time passing. I understand the patience of the tortoise over the agitation of the hare. Yoga forces you to slow down, to pay attention. I find myself open to receiving wisdom from other bodies that know more than mine.

That morning, there were thirty or so students in attendance. Some were in wheelchairs, a few were assisted by helpers, and a couple of the students were teachers or therapists themselves. We pressed our palms together in seated prayer pose, and closed our eyes.

Matthew is plainspoken, his vowels Midwestern, his manner direct and calm. "For those of you who can and stay in balance, push through the heel of your hands, to keep hands at midline," he said. "Lift your sternum, lift your rib cage—this is a clarifying shape. Drop your head, let your brain be subservient to the energy coming up through your chest. Feel the outline of your ribs. Breathe into the shape. Raise your head up."

Inhale, exhale, inhale, exhale. We opened our eyes.

Because the day's class was held online, the participants were geographically far-flung. As I looked at the faces on my screen, I was moved by this grid of focused intention, of people finding their shape, all across the country and beyond.

Matthew bounced in his wheelchair and smiled. He demonstrated lifting up off his seat on his hands, to create space in the spine and wake up the lower back and the legs. After observing his students' individual poses for a moment, he gently offered adjustments as needed.

"Renee, as you lean in one direction, make sure to push off the opposite side. I'm stretching this way, but I'm making sure I pay attention on the back side too. I want to feel the shape with my whole body."

I followed the sensation of space in my spine down my legs to my feet. I flexed and stretched my toes and relished how good that small intricate unfolding felt. With Matthew's direction, I paid closer attention to subtle changes, inside and out.

"It's the only body you ever get to have," he told us. "As you stretch, don't push through its resistance. Let it have its voice, and figure out how to become part of it."

NOT LONG AFTER that class, we sit on the backyard patio of his light-filled two-story home outside Minneapolis, under an expansive canopy of maples that, at more than a century old, are approaching the end of their lifespan. It's a moody early-June afternoon; as rain and thunder move across town, we listen to the specific music of leaves applauding. Matthew tells me he likes to let the trees fall as their time comes, to the sometime chagrin of his neighbors.

As a graduate student, he studied philosophy at the University of California, Santa Barbara; though he has an elliptical way of talking, he always returns to the palpable details of physical life: blood, sweat, tears; muscle, and bone.

We lose body awareness due to age, injury, illness, trauma—it's the violent condition of being alive.

"Disability can happen to anyone at any time," he says. "And it *will* happen to anyone who ages."

Sudden, or slow. I've never thought about physical disability in this way, but there is clarity in this framing. When I first asked if I could attend his adaptive yoga class, he'd welcomed me warmly, but emphasized that it was important to participate too. "We learn by doing, and feeling," he said.

Later, I realized that this situated me on the continuum of

ability and disability, connected not just to my own body, but to everyone else's too.

What does adaptive yoga look like? With the help of his partner and fellow teacher, Molly Bachman, Matthew demonstrates some of the ways that the principles of yoga can be applied to bodies of differing abilities. One of the more remarkable poses they show me is a modified handstand—or, as Matthew describes it, "an inversion that can be done without ever going upside down."

First, the traditional handstand, in which Molly and I kick up and balance on our hands against the wall: "So you remember what it feels like, what muscles it involves."

Then, the modified handstand. We reposition ourselves flat on our backs on the floor, bodies perpendicular to the base of the wall and our heads at arm's length from it. "Reach your arms up behind you, place your hands flat on the wall, and look where you're reaching," Matthew instructs.

The effect is startling: the stretch of the muscles along my arms, ribs, and abdominals, the arch of my back and neck, the feeling of reaching and reorienting the spine, the grounding of hands in the "earth"—in this case, the foundation was the wall—all culminate in the feeling of going upside down onto my hands, just with a little less gravity. It speaks to the flexibility of muscles *and* the brain to receive the benefits of yoga.

Matthew and Molly also show me a seated pose that their students call "running man," in which a partner pushes against each of your knees, alternately and in rhythm, to create a sensation akin to the way the hips move while walking and running.

After just a few pushes from Molly, the motion loosens

something in my lower back and hips, giving me a welcome kind of relief, and Matthew smiles when he reads the expression on my face.

"It creates a sense of lightness through your torso," he explains. "Your spine delivers the signals in a way that your mind isn't even necessarily aware of. But your brain is receiving it."

Later, he suggests that it might be something to try with my father-in-law, to give him the same relief. This is the wisdom of bodies as a continuum.

AT MY COMMUNITY pool, the locker room is a tableau on aging.

Throughout the day, you'll find bodies and bottoms of every type on display, ranging from squishy baby to saggy lady. But this is not the kind of place where short-lived resolutions to lose fifteen pounds get made or broken. Here, the arc of fitness is long, and it bends toward seniors.

The hour when I frequent the pool for my lap swim has long coincided with the 8 a.m. aqua aerobics class, taught by Kathe Rothacher, a calm, convivial woman with honey-colored hair and a beatific smile. Many of her devotees are in their eighties. Some are there for physical therapy after an injury; others are contending with the incessant aches and pains of age. In that crowded warren of benches and communal showers, where every flick of a towel or reach of an arm brings you into someone else's personal space, ordinary civilities carry larger import.

I am not eighty. But among these eighty-year-olds is where I like to be. I first came to this pool after my second child was born and my family moved across the bay from San Francisco to Berkeley. This is where I reclaimed my body, a little softer and a lot more tired, as my own. Day after day in the outdoor pool, I pulled and kicked my way back into the swimming habits that made me feel like, well, me.

A decade later now, my passage through many a day is eased by the morning transit through this locker room, in the company of these women. The daily celebration of bodies that are happy and working makes me comfortable and ever grateful in mine. Even when it's a stretch, we still try.

Here is where we warm up from the swim in the open showers; here is where we jockey for space in the cramped dressing area, all of us in various stages of nakedness—this one applying moisturizer, that one in underwear, still another wrestling with a stubborn pair of leggings. We contort our bodies in the most unattractive ways. It's where we can show vulnerability, in all its forms.

Loneliness, we know, leads to deteriorating health. I listen to the way the people in this room rally around one another, through struggles that range from family discord and sleeping woes to cancer and chemo and the death of dear ones. Sometimes I swim with a buddy or train with the Masters team. Often I come alone. But always I find company in the locker room—a conversation to dip into or just to listen to. And always there is the reassuring routine of simply discussing the water conditions in the pool that day or admiring the pattern on someone else's bathing suit.

Sometimes the struggles are mine, and sometimes they are the struggles of someone twice my age. I remember the day I was reduced to tears by the wildly swinging hormonal pendulums of breastfeeding and sleep deprivation. I remember missing my mother, who lives across the country in New York. That day, a woman told me that she was anxious about visiting her very pregnant daughter and how she wasn't sure how long her daughter wanted her around after the baby was born.

"Go," I told her firmly. "You can always ask her when she wants you to leave."

Certainly, there are maternal and grandmaternal surrogates to be found here. Once, as we were getting dressed, I confessed to a friend that I didn't know how to buy underwear anymore, because all of it comes from my mother, who can eyeball the ideal fit of a bikini brief for me from a mile away and who refreshes my collection of undergarments every year, in my Christmas stocking, without fail. I told my friend that my husband had expressed his incredulity to me in this way: "You're forty years old, and your mom still buys your underwear?"

But I remember that another woman, perhaps a decade older than we were, listened to the story and started to cry.

"That is the sweetest thing I ever heard," she said, wiping her eyes. "Tell your mom I said so." And I did.

I adore the cross section of ages and bodies in this locker room, but I like to observe the ecosystems of other locker rooms too. Not every locker room has the range of age and swimming ability that I now know to look for. There are plenty of sports clubs that attract impossibly fit Ironman triathletes and squash pros and disciples of the latest HIIT/cycling/barre/boot-camp workout craze.

But there's something vitally different about seeing older bodies as a younger person. As a society, we tend to fear the old and the aging and silo them away. Many of us don't often see our older relatives, much less see them naked. But there is an important kind of bond, a mutual acceptance and acknowledgment, that comes from having bared your flesh together.

I remember vividly a morning when Alicia showed Patricia her longtime stretching routine by getting right down on top of her towel on the clammy tiled floor.

"I have been stretching all my life. I have scoliosis," Alicia declared, mid-hip stretch. "If I didn't do it, I'd be in a wheelchair now."

Lovely Patricia with her British accent chirped anxiously above her: "I'm glad you're not! But I think you'd better get up now, dear, or you'll get run over."

Kathe has muscular dystrophy, which progressively affects the skeletal muscles of the body. She regularly dispenses nuggets about everything from mahjong to the history of Title IX at UC Berkeley. "I made this choice when I got diagnosed, to not *be* the diagnosis," she tells me one day. "I am me first."

For the past few decades, she has kept her body moving with the whole gamut of exercise, teaching everything from cycling and dance to yoga and Feldenkrais. She remembers learning about Matthew Sanford from her years as a yoga instructor at UC Berkeley. Like Matthew, she knows how important it is to learn how to reinhabit your body as it changes.

"With any trauma, you have to be able to come back into yourself," she says. "You have to be able to say 'That was what I used to do,' and then figure out ways to adapt to what's present—and what's useful. You have to always stay open to what's possible."

Every day in this locker room, I see people making the choice to do so. At every stage of life, I see that muscles matter.

As the ladies shuffle or saunter or sail out of the locker room, they call their farewells. When they pass me at the long mirror by the door, they smile and meet my eyes, and as they do so, I know there's a solution for most every problem in that small greeting.

How are you?

I'm all right. I'm here, aren't I?

Eighty-year-old women are full of wisecracks. Theirs is the kind of locker-room talk that teaches me to love the body I have today. To be open to the one I have tomorrow. Their laughs explode like a bouquet of fireworks, with a sharp and knowing joy.

12

It Comes from Unity

Ten years ago, at the height of her physical powers as a yoga teacher and aerial performer, Angelique Lele fell from a trapeze and lost the use of her legs.

Everyone who came to see her in the hospital—from doctors and nurses to family and friends—asked if she'd heard of Matthew Sanford. It got to be a darkly comic joke to her. *Enough already,* she thought. *If someone else mentions him, I'm going to scream.* Then the hospital chaplain came to visit.

"Have you heard of Matthew Sanford?" the chaplain asked. Before Angelique could answer, she went on: "Well, I asked him to call you."

Angelique began to cry.

"The first time we talked, Matthew practically taught me an entire yoga class," Angelique, a wry, funny redhead, with tattoos of butterflies flitting up and down her arms, tells me from her home in Florida. "I had already divorced myself from half my body. I didn't want to know what I couldn't feel. He immediately said, 'Can you feel your feet? Your sit bones? Just put your mind there.' It connected me to my whole body. It took me a long time to rebuild my love for it again, but that was the beginning."

As Angelique learned to do yoga as a paralyzed person, she tried to practice vinyasa the way she was accustomed to doing it, and got injured: a torn muscle in her thigh, a broken knee. Matthew told her, "You're just grieving, you know."

"He has a way of getting right to the point," Angelique says. "He was funny, and also very patient. He showed me how intense Iyengar could be—that my yoga practice could have the fire and intensity I missed from vinyasa flow, but with the stillness and beauty of Iyengar."

She started teaching her regular yoga classes again, and took a few training workshops with Matthew. But she was afraid to teach adaptive yoga herself—"afraid of taking other people's injuries in my hands," she explains. She watched Matthew teach, and saw how perceptive he was—how he would just recognize things in other people, and know how to reach them.

Over time, she softened. People in her regular classes had injuries, got older, and had to adjust how they did things. "One of my students had cancer, and his practice changed a lot," Angelique says. She realized that she was already figuring out how to teach her students where they were. "It was just that little leap that showed me that yoga is a way of connecting humanity. And to figure out with yoga how to meet the needs of people's individual bodies, so that it's not so hard."

Angelique now teaches three yoga classes a week for Matthew's nonprofit, Mind Body Solutions, which also offers regular trainings and workshops for yoga teachers, caregivers, and health-care professionals. For all that she has learned, she is quick to acknowledge that the journey has not been perfect, or easy.

"My body is changing all the time," she says. "All of our bodies are doing that on some level. But yoga helps you be still enough to notice that, and be at peace with it."

WHEN YOU STRETCH a muscle, the fibers are fully extended; if you increase tension even further, the connective tissue around the muscles—tendons, which attach muscle to bone, and ligaments, which attach bone to bone—also stretch. Stretching and moving helps to lubricate muscles and their fascia, clear waste products, and realign fibers from scarred muscle and connective tissue, which tend to be disorganized and clump together. The flexibility of fibers is important, in that it helps to support a full range of motion. But stretch too far too fast, and you'll cause damage to the muscle, and to that surrounding tissue.

Muscle spindles are our secret stretch detectors. They are receptors in our muscles that sense how much and how fast a muscle lengthens and shortens, and alert us to harm. If joints and connective tissues are in danger of being stretched too far, the spindles send a signal immediately to the spinal cord, which then sends a reflex response to contract the muscle further and protect against potential damage.

Signals like these are quieter or altered when you are paralyzed.

When I start digging into research journals, I discover that what Matthew talks about as presence, grounding, and reestablishing the boundaries of the body in yoga practice after spinal cord injuries has only recently begun to be described in scientific language. *Peripersonal space representation*: This is an unwieldy term, but it includes information gathered through proprioception and interoception, and visual and other bodily signals.

Proprioception, your body's ability to sense itself in space, is how you stay upright, balanced, and moving safely through the world without thinking about it too hard. Generally speaking, external receptors in your muscles, skin, and joints send messages to your brain. Interoception is your body's ability to sense itself from inside. It encompasses how your body *feels,* in terms of temperature, pain, pressure, hunger, fatigue, even your depth of breath. Internal receptors in your organs, bones, and blood, and also your muscles—remember those spindles—transmit molecular, biochemical, and electromagnetic information, often below the level of consciousness, to help the body maintain homeostasis.

Interoception can influence subconscious patterns of muscular tension or stress. Think about the way you might favor your right leg after an injury to your left ankle; this pattern is likely to persist long after the acute injury has healed and the noticeable pain has subsided. Instability and inflammation can take months or even years to resolve. Before you know it, that's your body's default way of doing things, to protect the "bad" ankle. Ironically, this self-protective pattern can lead to a chain of new injuries, which is why it can be important to identify and correct the underlying motor pattern before it causes more harm.

A functional sense of the body in space is diminished for someone who is paralyzed through injury or disease, because voluntary movement and surface sensation are limited. But a lot of awareness can be restored through the mobilization of affected limbs—as with, say, a yoga practice. A recent study with paraplegic and control subjects shows that motor feedback, not just visual cues, is critical to the recovery of healthy peripersonal space representation. Cognitive functions such as the representation of space, the researchers wrote, "are grounded in sensory-motor functions and bodily signals."

The science is so new that the mechanisms underlying this recovery are unclear. What *is* clear, however, is that moving the physical body itself, both actively and passively, helps to integrate signals for presence.

Matthew has spoken about his experiences in front of many audiences, including at the Mayo Clinic, where he was first treated for his injuries. Several years ago, a scientist at Rutgers University put Matthew's head in a functional magnetic resonance imaging (fMRI) machine to test the purported sensation he felt in his yoga poses. During one pose, in which researchers pushed on his feet, his sensory cortex lit up—and so did his motor cortex. In other words, his brain received a level of sensation from his feet, legs, and spine that translated to information for movement.

"Minute input and stimuli in my legs light up my motor cortex—it's what happens when I prepare for a yoga pose," Matthew tells me. His paralyzed body still transmits signals to his brain, he explains, and when everything is aligned, he knows it. Through his practice, he says, he has learned to listen interoceptively. His mission is to show others how to reconnect to this kind of subtle sensation in their bodies, with yoga.

I ask him: "What does yoga feel like when you're doing it right?"

"It's like the difference between a good hug and a bad hug," he says with a laugh. "We all know when it's a bad hug—when someone isn't connected with you. And you know when a hug is good. It's profound. It feels complete. It's integrative—you know where your body begins and ends." He pauses, thinking.

"A good hug won't cure cancer," he says at last. "But it will make the experience of living with cancer more bearable."

OUR BODIES ARE moving toward living all the time, even without our conscious control.

The muscles of the heart, blood vessels, and gut aren't under our voluntary purview, but they conspire to give life a steady rhythm, fast and slow, and, despite change, a continuity of being. As part of the autonomic nervous system, smooth and cardiac muscles carry on their patient, enduring work over the course of our lifetime—to move food and pump blood, say, or to generate heat, by raising the hairs on the skin.

Hair is one of the defining characteristics of mammals—even whales and dolphins have hair. At the base of each hair follicle there is a tiny muscle, the arrector pili. At the other end of that muscle is skin tissue. In Latin, *arrectus* means "raised," while *pili* means "hair"; when your body is cold, all those tiny muscles contract at once to warm you up, causing your hair to stand on end and giving skin the chicken-skin appearance we know as goose bumps. You can think of arrector pili as the ribbon tied around a bunch of flowers: The muscle holds a series of hair follicles in a unit such that when it contracts, it pulls them up to ninety degrees.

But cold is not the only thing that stimulates arrector pili contraction—fear and other stresses can fire up goose bumps through the release of adrenaline. Picture a porcupine's spines going up. The spines are actually long, thick hairs that help the animal seem larger and more intimidating to predators. Goose bumps can also be a response to an intense emotional state like awe. Imagine yourself listening to a stirring piece of music, watching a huge wave roll in from the horizon, or peering over the edge of the Grand Canyon.

Getting the chills: It's the body in a state of awe and wonder. It is a sign that we are trying to grasp the unknown. Every part of us, from arrector pili to heart and gut, is collectively engaged in making sense of the world. Even when we dream, new research shows, our muscles are twitching in REM sleep to improve the brain-body connection. We are learning and relearning our own bodies all the time.

The neuroscientist Mark Blumberg has found that not only do many animals twitch their muscles in REM sleep, but they start doing so before they're even born: First, the muscles move, and then the brain responds with bursts of activity in the sensorimotor cortex. He has watched slumbering octopuses flush with color as the muscles controlling their chromatophores contract; he has observed sleeping rat pups flick their paws and—moments later—the response of correlating motor neurons in the rats' brains, recorded by electrodes. The brain is listening to the body, learning in a granular way which motor neurons control which muscles, one at a time.

Blumberg and others have gathered evidence that this process of practicing while sleeping continues throughout our lives, as our bodies change from the womb through birth, puberty, and older age, through growth, injury, illness, and the learning of new skills.

The act of distinguishing different body parts is also beneficial in the practice of yoga, Matthew explains. "Even for a student who is paralyzed," he says, "something emerges from that."

In Iyengar yoga, a single pose can hold all the teachings of yoga. To lie in savasana, or corpse pose, is to focus on the sensation of being grounded, and to integrate all the poses that came before. People think about all kinds of acrobatics as being yoga. But something as quiet as savasana—that's a yoga pose too. When you go part by part—whether in the study of anatomy or in the practice of yoga—the whole can be illuminated.

FOR MANY YEARS, Matthew taught a weekly yoga class at the Courage Center, a rehabilitation center for patients recovering from profound mind-body injuries. One day, he worked with a young gunshot victim, a somber teenager who had recently lost the use of his legs.

"I was trying to help him be in his body again," Matthew says. "This kid was going through a lot, and he needed to find continuity. You can't go back, but you need to find out how to be whole. I wanted to teach him how to rise and lift himself to transfer out of his wheelchair." Matthew knew that the boy used to play basketball, so he asked him to remember the feeling of coming up for a jump shot.

"He was paraplegic, but he had an intuitive sense of directionality," Matthew says. "He could glimpse who he was. He was like, 'Holy shit. You mean I still get to have my legs?' The next day, he was getting fitted for a new chair cushion, and he said, 'I'm taking this one, because it's good for my jump shot.'"

That's the physical continuity that helps you to feel whole. Matthew argues for a reestablishment of that connection—not through the miraculous healing of the spinal cord or the curing of a muscle-wasting disease, but in reintegrating an awareness of the entire body, past and present, which has been fractured.

Before he found yoga, Matthew says, he used to think of

himself as a floating upper torso. Now he moves with his whole body. In conversation, he is fully animated in a way that integrates every part: leaning forward to make eye contact, sitting up sharply, using his hands to cross one leg over the other, raising his arms in the air to emphasize the point.

"I practice yoga not to just be really great at yoga poses," he says with characteristic frankness. "I practice yoga to feel good. To feel like a whole being."

There's magic between the muscles and the bone—this is the language of movement. After a great swim or a surf, there's a hum in my body; I know that I've moved right and that I've spent this time well.

"The fatigue is also cleansing, because you've lived through your whole body," Matthew tells me. "That sensation may not be the same for you as for me, because I can't swim a long distance or surf a big wave, but the energy is still there. The energy in the spine seeks expression, and our muscles and limbs are the main vehicles for that."

I think of the sensation that comes from being confined to bed, or to a cramped airplane seat during a long flight—a restless electrical energy that starts as an itch in the spine and builds and extends to the whole body when there's nowhere for it to go. Or the frustration of a foot injury—it's as if the foot is a flat tire, but the engine of the spine is still humming and grinding, throwing off sparks.

Matthew explains that with adaptive yoga, if you put your foot on the sacrum of someone who doesn't have a leg, that energy will go out the missing limb. "The alignment gives the energy direction," he says, "and that feels good."

We talk jump shots and waves surfed, but what we're really talking about is a feeling of completeness and connectedness with the world.

"My life force isn't completely determined by the ability to flex muscles," Matthew once told the interviewer Krista Tippett. "I've always felt that surge. I also know that that connection was what made me such a good athlete as a little kid. It's like, you feel a free throw. And it comes from your legs, and it comes from your arms, and it comes from unity."

He resists the perception that he is some kind of monk, or someone who is superdisciplined in his practice. "I struggle, I mess up, just like everyone else," he tells me, adding that dealing with the constant challenges of a body that doesn't behave can be exhausting. "But what makes me a damn good yogi is that I'm always willing to start again. That's the difference."

IN THE POEM "A Suit or a Suitcase," Maggie Smith writes about the body at the end of life. Do we wear the body, or does it carry us? What if what I think of as *me* were distributed everywhere in that body, thoughts and feelings and the sense of self in the hands and feet as much as the head and chest?

"Ideas are whispering in my wrists / and all along the slopes of my calves," one stanza goes. What if we were equally present in all of those moving parts? What would they tell us about what we miss when we don't acknowledge them?

A few weeks after visiting Matthew, I find myself walking by a yoga studio in Tokyo. I think about continuity. On a whim, I sign up for a week's worth of classes.

The studio is small and intimate, and in my first class, we are seven bodies moving together. There are no mirrors here;

no constant reminder to care how I look to others. The teacher speaks almost exclusively in Japanese, so my mind is set loose, free to observe, as my body does the work of moving through the poses. The motor neurons that know yoga flow take over, and my muscles know what to do. *Mountain pose. Forward fold, flat back, jump feet back to plank, chaturanga, upward-facing dog, downward-facing dog. Breathe.*

The hour passes by in a quiet stream of Japanese. Then, near the end of class, a few words in English: "Feel your body. Be here now."

Countless nows have passed, but I hold them all in this body. Pay attention to fingers and toes, the little muscles pulling those levers; be aware of eyes and neck, everything working in tandem. The shiver traveling up the spine, the arrector pili raising goose bumps in its wake. I find myself thinking about, of all things, death. Yoga won't solve mortality. But the way it connects me to all the parts of my body—and to a greater understanding of things I don't like thinking about—helps me be more accepting of what's to come.

MY FATHER-IN-LAW, THRIFTY with words, is one to hold his cards close. But ask him about your golf swing ("Keep that right elbow straight!"), or what it felt like to run a leg of the 4 × 400-meter relay at New York's Madison Square Garden under the coaching of his dad, Jumbo ("Terrifying"), and suddenly he is spirited, full of opinions.

As a high schooler, he was drafted by his dad to practice with the Villanova track team in the afternoons. He shared track time with some of the greats, including the distance runner Dave Patrick, who would grace the cover of *Sports Illustrated*, and

Frank Murphy, a two-time Olympian from Ireland who won multiple NCAA championship relay and team titles.

It's late summer, and we've come east from California to visit Jim at his home in Pennsylvania. The boys are whacking golf balls from the driveway across the front yard; we watch them from under the shade of the sycamore tree, Jim in his wheelchair and Matt and me on the grass. Jim tells us about running for his dad freshman year at Villanova, and about how he decided to leave track behind for a lifetime of golf.

Jim, to his father, with trepidation, after a disappointing showing at Madison Square Garden: "Dad, I think I'm going to play golf now."

Jumbo, dryly: "I think that's a good idea."

A lightbulb goes on in my head. I ask Jim if I can show him the running man pose that Matthew Sanford taught me.

"Sure," he says. He sits up as straight as he can in his wheelchair, and I place my hands on his knees, taking care to plant my feet before we begin. As I push firmly on his knees with my hands—first one knee, and then the other—I explain how the alternating motion can help his spine and hips release their energy through me.

Teddy, ever watchful, asks what we're doing.

"I'm taking a walk with Grandpa," I say.

I'm his muscles, grounding his body with my own two feet. The feeling of movement on a foundation of physical stability and solidity, connected both to the earth and to him, is how I understand that you can help someone else regulate their body by using your own. There is a rare intimacy in this act.

Before my father-in-law thanks me with his words, his face

tells me there is relief. To me, this is what it means to put muscular action in service to the whole.

A couple of mornings later, I ask Jim if he'd like to take a walk with me, and he brightens. We do the running man, right there in the kitchen.

13

Remembrance of Exercises Past

When I interview a longtime sports medicine doctor at UC Berkeley named Harris Masket, whose job it is to work with athletes playing dozens of intercollegiate NCAA and club sports, I ask him what he finds most fascinating about muscle. All day long, Harris ministers to a parade of athletes of various shapes and sizes, with bodies that do all kinds of extraordinary things.

"They'd walk in, and I'd think 'You play volleyball,'" he tells me. "Or 'Oh, you're a swimmer,' 'You do gymnastics,' 'Oh, you're lacrosse.' The body and muscle types are so different, and it's amazing how the muscles work for each particular person."

They'd ended up in his urgent care clinic with the whole gamut of wild muscle-related injuries, which told mind-bending little stories in and of themselves: myositis ossificans, in which bone forms in injured muscle tissue (muscles have incredible blood flow, which helps them heal, but sometimes calcium gets into the tissue architecture as it is laid down during the healing process); rhabdomyolysis, in which muscle tissue breaks down and releases proteins into the blood, which can result in kidney failure (rhabdo is nicknamed CrossFit disease, since it's often caused by intense overexercise);

and unusual musculoskeletal conditions like Lisfranc injury, which at first seems like a minor midfoot sprain but, in fact, without surgical intervention critically hinders walking ability (it's named for a French surgeon who served in Napoleon's army and observed that patients suffering from this injury could no longer be effective marching soldiers).

"What I really enjoy is the fast-twitch muscle athletes who say 'I've been playing rugby forever, but I'm twenty-two and I need to get out of contact sports. What else can I do?' And then they start doing triathlon," Harris tells me. "People who say 'All right, can I change what I do with what I have?' And then they excel in the new thing, and become someone else. I see a lot of athletes who have overcome injuries and other challenges. That really ties into malleability, change, and personal power."

What he loves most is the tissue's penchant to be flexible, no matter what happens to it. It's in its character.

"Why do we like muscle?" he says. "Muscle is the ability to change."

WE ALL WANT to know if and how we can come back to form after injury, illness, or a long hiatus. Muscles adapt in response to the environment: They grow when we put in the work, and shrink when we stop. But what if we could help them remember how to grow?

As a general rule, cell biologists don't enter their careers by running through the gauntlet of top-tier professional sports. But in the years that Adam Sharples played as a front-row forward in the UK's Rugby Football League, he found himself wondering about cell mechanisms that helped muscles to grow after different types of exercise.

A front-row position in pro rugby means that you have to be, well, "quite big," as Adam puts it. "I was in the gym lifting weights from the age of about twelve, I think," he says.

He spent much of his teenage life in training. When he was nineteen, he was playing a Boxing Day match on soggy ground that was heavy underfoot. He'd just planted his foot when a player on the opposing team tackled him, torquing his upper body to the left. His right foot remained firmly stuck in the mud.

"That's when I tore my ACL, but I don't remember much about it. You should ask my dad," Adam tells me with a wry smile. "He could tell you down to the minute, in great detail: when it happened, how it happened."

When I hear this, I'm moved once again by how sports has the remarkable capacity to be a love language.

Adam took a year off from rugby and continued to study, completing his master's degree in human physiology. He'd always been curious about muscles and muscle growth, but the hiatus gave him time to think—pro rugby players, he was well aware, have notoriously short careers. That acknowledgment eventually led him to pursue a PhD in muscle cell biology.

When we talk about muscle memory, most of the time we refer to the way our bodies seem to remember how to do things that we haven't done in some time—riding a bike, say, or a complicated dance we learned in childhood. When you learn and repeat certain movements over time, that movement pattern becomes refined and regular, and so does the firing pattern of neurons that control that movement. The memory of how to perform that action lives in our motor neurons, not in the actual muscles that are involved. But as Adam proceeded through his academic training, he became more and more interested in the question of whether muscle itself possesses a memory at the cellular and genetic level.

Almost two decades later, Adam teaches and runs a lab at the Norwegian School of Sport Sciences in Oslo. In 2018, his research group was the first in the world to show that human skeletal muscle possesses an epigenetic memory of muscle growth after exercise.

Epigenetic refers to changes in gene expression that are caused by behavior and environment. The genes themselves aren't changed, but the way they work is. When you lift weights, for instance, small molecules called methyl groups detach from the outside of specific genes, making them more likely to turn on and produce proteins that affect muscle growth. Those changes persist; if you start lifting weights again, you'll add muscle mass more quickly than before. In other words, your muscles remember how to do it: They have a lasting molecular memory of past exercise that makes them primed to respond to exercise, even after a monthslong pause. (*Cellular* muscle memory, on the other hand, works a little differently than epigenetic muscle memory. Remember how exercise stimulates muscle stem cells to contribute their nuclei to muscle growth and repair? Cellular muscle memory is when those nuclei stick around for a while in the muscle fibers—even after periods of inactivity—and help accelerate the return to growth once you start training again.)

Athletes have always known this to be true, at least anecdotally. After periods of injury, as with a torn ACL, they notice that it's fairly easy to regain the muscle strength they lost. The joints, though, are another story.

Adam took his reconstructed knee and ground through another year of pro rugby before retiring for good. In his academic work, he began to investigate the *why* behind his observations about muscle memory. In doing so, he found a way to grapple with what it means to age as an athlete, and as a human.

"Looking back, I was probably overtraining in the attempt to be the best I could be," Adam says. "Because if you can find the exercise that provides your muscle with the longest-lasting memory, or find the type of training that your muscle can respond better to the second time around—after an injury, say, or after taking some time off—then you can potentially reduce the amount of exercise you do for the same benefit."

He laughs. "I could have saved myself some work, I suppose. I've got that hindsight now."

HOW DOES EXERCISE over time influence our muscle memory? Can we alter our exercise training to lead to a memory of that prior exercise and have our muscles adapt? This is where the work of Adam's muscle lab—and that of others researching cellular and epigenetic muscle memory—has seized the public imagination. The appeal is obvious for athletes who are looking to boost muscle mass and strength. The research also has far-reaching implications for steroid ban policies in competitive sports, since the changes the drugs create may confer longer-lasting advantages for muscle growth—months longer, or perhaps years longer—than previously thought.

But Adam's research is even more relevant to an aging population—which is, well, *all* of us. Can we use that adaptation to protect ourselves as we get older?

The Norwegian government hopes so. Within the next decade, Norway is expected to become a "super-aged society," in which the share of the population sixty-five and older exceeds 21 percent. The country's health agency funds his lab's investigation of whether human skeletal muscle remembers not just periods of muscle growth, but also whether muscle remembers periods of *wasting*—and what to do about it.

As we age, our muscles lose strength and mass—a normal process known as sarcopenia. This gradual loss of skeletal muscle generally begins sometime in our thirties (surprise!) and hits the accelerator by the time we get to our midsixties. Like most everything else in our bodies with the passage of time, muscle cells do their work less effectively—with fewer and shorter fibers that are less able to process energy—with a corresponding decline in the mitochondria that power them, the motor neurons that control them, and the hormones that regulate muscle growth and atrophy. Sarcopenia happens to everyone, even to active athletes, but it happens much faster if you don't move much; the age-related loss of muscle is a major cause of frailty, increasing the risk of falling and fractures in older people.

Menopausal women are especially vulnerable to sarcopenia. The loss of muscle correlates with dropping levels of estrogen, which normally performs the important roles of stimulating muscle satellite cells to repair tissue and helping to regulate metabolic and mitochondrial function. Though women generally have less muscle and lower aerobic capacity than men, women need to do less exercise than men to get the same benefits in longevity. And stronger skeletal muscle influences other muscle: New research also shows that women who do weight training just a couple of days a week reduce their risk of death from heart disease by a whopping 30 percent.

Strength training is the single most important thing you can do to slow or even reverse the normal muscular decline of aging.

But people don't always receive this prescription for a daily dose of iron—pumping iron, that is—from their doctor, or take it seriously enough to fill it when they do.

Globally, Scandinavian countries—Norway, Sweden, and Denmark—have the highest incidence of hip fractures due to falling, especially in (and because of) wintry weather.

"Repeated falling in the elderly is a real issue. People fall over, they lose muscle, they become more frail," Adam says. "Some of them might recover, but not properly. And then they might fall again, and lose more muscle. And that's linked with early morbidity and mortality—they get sick and die earlier."

Falls are the second-leading cause of unintentional death worldwide, and muscle weakness is a major risk factor. In Norway, hip fracture caused by accidental falling is the strongest predictor of unexplained death in older people. In fact, falling is so common that there are public-service commercials on Norwegian television for how to fall properly. How *does* one fall properly? On your side with your hands out, as if you were getting your leg swept in judo. In other words, you want to avoid the banana-peel fall onto your back and head.

In the Netherlands, there are even standardized classes on how to fall—and how *not* to fall—offered to the elderly, in which participants navigate obstacle courses like "the Belgian sidewalk" and practice falling onto foam mats. (The name for this obstacle course may or may not be a Dutch joke about Belgian construction being inferior.) Hundreds of these falling classes are offered across the country, and they are so popular that the Dutch government rates them and health insurance covers them.

What Adam teaches me about muscle and aging feels especially poignant. A few years ago, my grandmother died after a fall down the stairs at home, a week away from her ninety-fifth birthday. Instead of celebrating her birthday with her, we mourned without her at her funeral. More recently, my mother has developed advanced osteoporosis, which puts her at serious fall risk too. She loves line dancing and walks a few miles a day; I beg her to throw some hand weights into her exercise routine. I tell her about a woman in her seventies who changed her life with weightlifting; in three years, she went from struggling to go up the stairs and suffering from high blood pressure, high cholesterol, and kidney disease to bench-pressing ninety pounds and being medication-free.

My mom says she doesn't need to be the Hulk—but adds that she'll try.

Something as simple as a ten-second balance test on one leg is a powerful indicator of mortality risk. Or the sit-to-stand test, in which you see how many times you can stand up and sit down in a chair in one minute, without using your hands. It's another way of testing balance through the strength of muscles in your legs, glutes, feet, and core. The stronger they are, the better your balance.

Muscles get older, but they can still be reminded to do what they did in their youth. They can be rejuvenated in the most surprising of ways.

"What we've found with our earlier research is that one of the genes turned on during initial muscle training seems to be one of the most strongly linked to the return of muscle mass after detraining," Adam says. "So if an older person falls, we might be able to intervene during the first recovery with a gene therapy injection in the muscle. What we're trying to find out with our current research is, does it increase the muscle recovery? And does it help when the muscle then encounters wasting later? Does it protect it and stop it from wasting more? What if we could provide a really important therapy to protect an aging population?"

I remember what Matthew Sanford told me: that every day we are seeking continuity. What if we could help our muscles remember?

ADAM SHARPLES LIVES with his wife and two children outside Oslo, in the old mining town of Kongsberg. He no longer spends hours at a time in the gym; instead, he gets up at 5 a.m. for an efficient session of weights before returning home to help get his young sons ready for school. He proudly shows me videos of the children learning to ski on the tree-lined slopes five minutes from his home. Adam doesn't ski, but he retains the broad, stocky build of the professional athlete he once was.

His visible bulk is a matter of gentle humor in his lab. "Adam just has to look at a weight and he packs on muscle," jokes Daniel Turner, the lead researcher on the muscle-wasting study. Daniel himself is a soccer player with the trim, muscular build of an endurance athlete. "Me, I'm all slow-twitch muscle fibers, so I spend most of my time doing resistance training hoping to put on even a little bit of mass."

I'd flown from San Francisco to Oslo in the middle of a winter storm, arriving just in time to meet Adam and Daniel bright and early on the sprawling campus of the Norwegian School of Sport Sciences. (Actually, it's dark and early, being the dead of winter in Norway.) Outside, fat snowflakes swirl down; inside the climate-controlled lab, Daniel introduces me to an affable, bearded man named Magne Lund-Hansen, who lies flat on his back in a body-scanning device, pretending to be injured.

Since it's unethical to experiment with repeated atrophy on older humans, the study recruited young, healthy humans like Magne, who is thirty-one, to hobble around on crutches and knee braces to induce muscle wasting in one leg. This limb-immobilization technique has also been used for space research; much of what happens to astronauts' muscle and bone in zero gravity approximates the effects of aging.

To start, Magne spent two weeks with his left leg immobilized in a knee brace locked at thirty degrees, to simulate what might happen after a fall. After coming in for testing—to see how much muscle strength and mass was lost in the affected thigh, and to have a muscle biopsy taken to see if epigenetic marks in the DNA were preserved—he was allowed six to eight weeks of recovery, followed by another two weeks of limb immobilization.

This morning marks the end of that second period of immobilization for Magne, and it shows: His left thigh is visibly smaller than his right.

"Let's see how much Magne has lost in muscle mass compared to the first time," Daniel says, eyes on his computer screen. He crunches data from a DEXA scan, which measures separate changes across muscle, fat, and bone tissue.

After a moment, Daniel has a number: "Looks like a little more than last time—about two hundred grams."

In response to my blank expression, he adds helpfully, "That's like, um . . . one to one and a half chicken breasts?"

So what does the loss of one and a half chicken breasts' worth of muscle mass mean, on a practical level?

Magne has some thoughts on the subject. "Well, the first time I tried to walk after limb immobilization, I almost fell down the stairs," he says dryly. An avid skier and athlete, he tells me he has a newfound appreciation for what it is to be an able-bodied person in the world. "It really surprised me, how slow my muscles were to respond to what I expected them to do."

We walk carefully down the hall to another lab testing room to meet Adam, who is rushing in late from his morning commute. "Wintertime in Norway," Adam says by way of explanation, a bit breathless as we shake hands. He'd roused himself from bed at 4:30 a.m. to dig his car out of the snow.

The snow, it turns out, is an important character in this story. I can't help but notice the array of footwear littering the floor of each researcher's office: shoes with spikes, heavy-duty hiking boots, nubby-soled sneakers for trail running, cross-country ski boots.

"There is a vast range of winter precipitation-related ground conditions in this part of the world," Adam says. "I've never owned so many different pairs of shoes in my life."

On crutches, as Magne learned, the snow becomes especially hazardous. Instead of his usual twenty-minute bike commute, it took him over an hour to get to work on public transportation that can accommodate his crutches and his knee brace. "That's quite a big difference each way," Magne says. "And it was tougher the second time around, because of the snow on the ground. The best time of day was coming home, taking off the knee brace, and getting on the couch."

The small, everyday tasks he took for granted were where he felt his immobility the most. "I bought a coffee for my girlfriend at a coffee shop and I didn't get ten meters before I dropped it," he says. "Once, I brought a lunch of pasta back to my office and sat down, and then I realized I forgot a fork. Rather than dealing with going back, I just ate it with my hands."

In the lab, a research colleague efficiently performs a biopsy on Magne's quadriceps muscle, extracting a small amount of cherry-bright tissue and depositing it on a tray. I thought of the anatomy lab: meat on the table.

Magne was surprised by how quickly he regained muscle mass between immobilizations. "That was the easiest part," he says. "But it takes longer to get everything else back—the motor coordination has to catch up to the new state of your body, whether it's that you're weaker or stronger. There's a lag time with the brain, and the expectation of what your body can and can't do."

This is Adam's theory in action: It's clear how someone who fell once and thinks they've recovered can easily fall again.

The human study was running concurrently with a similar animal study on aged rats, done in partnership with colleagues at Liverpool John Moores University in the United Kingdom and the University of Iowa in the United States. Adam and his team have identified the fact that the gene called UBR5 affects muscle size and recovery after a loss of muscle mass. In the rat trials, researchers used a gene therapy treatment to turn on UBR5 after the first period of muscle wasting, i.e., during the recovery phase. After the second period of muscle wasting, they would study the muscle to see if the treatment, in combination with exercise, could help prevent the repeated loss of muscle mass.

So far, they've found evidence to support their hypothesis that both young and aged muscles have memory of earlier atrophy, and that some genes are turned off when muscle is lost—the opposite epigenetic profile from what occurs with hypertrophy and repeated exercise, when genes are turned *on*. In the next several months and beyond, Adam hopes to identify more genes that have an epigenetic memory of muscle loss.

Scientists are also studying the ways other species in the animal kingdom have built-in protections against muscle atrophy. I'm fascinated to learn that serum from the blood of hibernating black bears not only prevents human muscle cells from atrophy when added to those cells in lab culture, but also *boosts* mass in those same cells.

While humans begin losing muscle mass within weeks of physical inactivity—leading to all kinds of cascading physical and metabolic dysfunction—hibernating bears can stay still for more than six months without eating or drinking, and without losing muscle mass and strength. After a summer and fall packing on calories, bears endure the winter without moving their muscles. Despite this lack of use, something in the bears' blood helps them resist disease and maintain those muscles for when they rouse themselves in the spring.

While bear serum for humans is still beyond the horizon, the fact remains that we *can* inoculate ourselves against the ailments of age by moving as much as we can.

My father was flummoxed when his father died. Understandably, he always feared that he would die young too. Since then, he has performed his daily exertions to build muscle—punching, kicking, jumping, and lifting his way to health—all in order to avoid that fate. Among my peers, he is the paragon of parental health, the model of endurance and longevity. At seventy-seven, he could be the model for an exercise campaign about how strong muscles can keep someone healthy and fit into later life—and looking a good fifteen years younger than their chronological age.

Lifelong exercisers in their seventies have cardiovascular capacity similar to active people many decades younger, says Scott Trappe, who studies muscle and aging at Ball State University's Human Performance Laboratory. Their muscles have enzymes involved with aerobic metabolism that are the same as exercisers in their twenties. And exercisers in general have more muscle capillaries, more supple, elastic arteries, and better blood flow to supply muscles with nutrients and oxygen.

Of course, we're always looking for a shortcut. Newly popular techniques like blood flow restriction, which uses tourniquet-like bands or blood-pressure cuffs on arms and legs during exercise, are employed by both endurance athletes and centenarians, to trick the body into speeding up the normal process of repairing and rebuilding muscle tissue. First used in Japan, blood flow restriction helps athletes reduce the training load, hours, and risk of repetitive stress injury, and helps older people build more muscle faster while using less weight. NASA researchers are even sending equipment into space to test whether astronauts can use the technique to build muscle and reduce bone loss in zero gravity.

"With resistance training in aged muscle," Adam says, "we've found that you can reset to some level the epigenetic changes to be on par with young people's muscle." Even exercisers who don't start until late in life can gain remarkable mass and strength.

No matter how old we get, the capacity to bounce back is there for us all.

IN THE AFTERNOON, after the battery of testing is complete, Adam, Daniel, and I head out into the snow for a brisk hike around Lake Sognsvann. As we walk the few miles around the lake in the cool gray winter light, our feet crunching and sliding every which way, we discuss what the flexibility of muscle has to teach us about how to live.

"Muscle is one of the very few tissues that adapts and changes on an everyday basis," Daniel says, peering at me through glasses wet with droplets. As cell biologists *and* exercise scientists, he and Adam are delightfully nerdy in their fascination with muscle and its elastic superpowers: the incredible ability to repair and regenerate, and how training coaxes muscle to increase even faster in size, strength, and endurance.

Muscles have become all the rage when it comes to chasing the fountain of youth, and understandably so. Scientists have recently discovered that with repeated exercise, immune T cells *in the muscles themselves* are controlling inflammation, maintaining muscle health, and boosting endurance. Stronger leg muscles are linked to better recovery after heart attacks. Even grip strength—how strong are the muscles in your hands?—is a good indicator of cardiovascular health and longevity; it's also positively associated with cognitive function. Muscles are also involved with basic metabolic processes, like regulating the uptake of glucose, which is why exercise is so important for diabetics; people with obesity have less muscle to aid in that process.

And muscle health is an accurate reflection of overall health. As Adam's longtime mentor Claire Stewart has demonstrated, muscle cells can carry the memory of impaired function associated with diseases like cancer, sometimes for years after the fact. In 2023, Adam's research group showed that muscle taken from breast cancer survivors even a decade after treatment shows an epigenetic profile similar to muscle of advanced age. But get this: After five months of aerobic exercise training, participants were able to *reset the epigenetic profile of their muscle* back toward that of muscle seen in a healthy, age-matched control group of women.

I remember what the dancer Ebony Ingram told me, after recovering from cancer: *I still need to jump.*

All the time I've spent with strongpeople and sports historians;

artists and anatomists; scientists and science historians; jumpers and surfers and yogis and runners: What all these people have shown me is that work—real, corporeal, physical *work*—is a critical and literal part of pushing back against loss. My own muscles are not outrageous, and they don't jump out at you with price tags attached. But I know what they're worth. I don't need to look like a strongman. And neither do you.

"WE ALL OWN a body, and we all perform life." This is something that Kevyn Dean, the medical director for USA Surfing, tells me as we observe patients in his physiotherapy clinic, and I'll never forget it. He treats Olympians and retirees with the same attentiveness to muscle and range of movement. "Athletes are my love, but so are the eighty-year-olds who want to be able to get a can off the high shelf," he says.

Muscles adapt and flex with us—as we grow and recover, as we switch our sports, as we get older. "No matter who we're training or treating, it's the same underlying principle of developing strength and control," Kevyn explains. The work doesn't end as you age, he says—in fact, muscle becomes even more important. "This is the work that builds endurance, and longevity."

And so I've been lifting weights. Doing push-ups, jumping rope, practicing yoga. Stretching and rolling my muscles. Joining my friend for morning boot camp, swimming laps in the pool, running intervals on the track. Trying to be an athlete at life.

ENDURANCE

. . . to stretch, heave, grimace if need be
as the finish line approached.

—MARGO JEFFERSON

14

What We Carry

Perhaps the single feat that most influences how we think about human running endurance today dates back to the year 490 BCE, when the Persian army landed on a plain in eastern Attica, en route to attack the city of Athens.

Situated on this large, fertile plain was the village of Marathon, which translates from the Greek as "fennel field." The Greek writer and historian Herodotus sang the praises of a long-distance courier named Pheidippides, who, when the news of the landing reached Athens, ran to Sparta with a request for help. To get there, he had to cover a distance of more than 150 miles over grueling, mountainous terrain.

As Herodotus told it, Pheidippides arrived in Sparta the very next day.

Though the Spartans couldn't get to the field in time to join the fight, the allied Greek forces managed to defeat the much larger Persian army at what became known as the Battle of Marathon. The Athenians then marched back to their city and successfully repelled further invasion.

One variation of the Pheidippides story also has our hero running from Marathon to Athens with news of the victory; this is the version that inspired the Robert Browning poem "Pheidippides" ("So, when Persia was dust, all cried 'To Akropolis! / Run, Pheidippides, one race more!'"). That, in turn, inspired our old friend Baron Pierre de Coubertin to introduce the marathon race at the first modern Olympics in 1896. And it's why early marathon events varied in length around 25 miles—it was based on the approximate distance between Marathon and Athens. The 26.2-mile marathon made its debut at the 1908 Olympics in London, and became standard soon after.

Note that this version of the story also has Pheidippides dropping dead upon delivering his message ("Joy in his blood bursting his heart, he died—the bliss!"). Based on the original Herodotus texts, historians deem this additional leg unlikely to have occurred, but the point was made: The guy was committed to the bitter end.

Two and a half millennia—and billions of pairs of Nikes—later, we have come to use the word *marathon* broadly, to describe just about any long event or trial. The entirety of the word isn't even necessary to conjure its larger meaning; the extracted suffix *-athon* is sufficient to indicate a prolonged bout of activity—a sport other than running (say, swimathon, bikeathon), but also completely sedentary pastimes (readathon, talkathon). It's a stand-in for endurance itself.

Despite the winding path the marathon has taken to get to us, what has remained from the start is the idea that it is a test of tireless determination—and that some noble measure of struggle and discomfort is baked into the endeavor.

THE STRUCTURE AND physiology of the human body reflect that, in many ways, it has evolved to run far and long, efficiently and without injury. Many of us are familiar with the contours of the origin story of our species: how we descended from ancestral hominids that had adapted to life on flat terrain, traveling long distances while foraging and persistence hunting; how we became exceptional at chasing prey until it collapsed from exhaustion, often in the heat of the day. Adaptations that aided in this pursuit include a profusion of sweat glands, to prevent overheating; short toes, longitudinal foot arches, and an elongated Achilles tendon, for efficient propulsion and flexibility; and a high percentage of slow-twitch muscle fibers, which resist fatigue.

As Christopher McDougall writes in *Born to Run*, the human body is able to respond dramatically to physical demands, quickly revving up everything from heart rate to mitochondrial activity and red blood cell count to deliver energy where it is needed: the muscles. The heart pumps a steady supply of nutrients and oxygen to the legs; during exercise, blood flow to skeletal muscles can increase *up to a hundred times* above where it is at rest. Pacing is important for endurance, because moderate speed and intensity helps to preserve muscle tissue by burning fat, which slows down glycogen depletion and delays fatigue. Sprint at the start of a marathon, and you will hit the wall sooner rather than later; push far enough, and muscle itself starts to be consumed as fuel. A combination of running and walking is common in persistence hunts as well as modern ultramarathons, to conserve energy. In the longer term, our bodies are able to reshape themselves to match continued demand—or lack thereof. Muscle has great capacity to keep doing the work, provided that conditions require it.

And if endurance is vital to the story of human evolution, one muscle in particular plays a starring role.

The gluteus maximus is the biggest muscle in the human body, and modern humans are the only animals on Earth that have such a large gluteal muscle. As science writer Heather Radke put it, "to succeed on the savanna, large gluteal muscles were a must." It's a uniquely human feature, one that is relaxed during walking but fires up and clenches during running.

Biomechanically speaking, the Harvard anthropologist and evolutionary biologist Daniel Lieberman explains, running is really a kind of controlled falling. The glutes are integral because they allow for leg extension and stabilize the body upon landing, so that we can run steadily for a long time without hurting ourselves.

At least one scientist has called the gluteus maximus "a multifunction Swiss Army knife"—not only does it make us efficient at long-haul running, but it also helps us perform other movements essential to our survival, among them lifting, squatting, climbing, throwing, and jumping.

In an era when the gluteus maximus may well be used more for sitting than as the engine of propulsion, running a very long way takes on added resonance. The act is a test of both physical endurance and the mental fortitude to continue in the face of pain and fatigue.

The sports journalist Alex Hutchinson, who has written extensively on the limits of human performance, points out that "the will to endure can't be reliably tied to any single physiological variable." The mind plays a critical role too.

In the late stages of prolonged exercise, when your muscles and your liver have exhausted their glycogen stores, your blood glucose level drops—this is the "bonk," as your brain and muscles register the signal that your fuel tank is empty. Your body begins to break down muscle itself for energy, converting its amino acids to glucose. This is when making the decision to stop presents itself.

Choosing to keep going, despite all signs telling you to stop, is at the core of the effort. So why do it?

Running with purpose is critical, as Lieberman and a research team of anthropologists explained in a recent paper about humans and long-distance running; to Indigenous societies like the Tarahumara (also called the Rarámuri) of northwestern Mexico, endurance running is a form of prayer and of forging social ties within and among communities—this applies to team footraces as well as hunts, and in relationship to other physical activities that require endurance, like dancing and farming.

We are deeply social animals. There are practical, social, and spiritual reasons to run, and researchers observe that these same elements are common in major marathons all around the world.

What I'm trying to get at here goes beyond caloric expenditure, refueling strategies, and what happens inside a muscle when it's utterly spent. What moves me most is the ineffable catalytic stuff of muscling through.

Attempting a marathon—or, for that matter, an ultramarathon, given that Pheidippides may have run more than 150 miles—isn't just about your muscles and physical abilities going the distance. It can be shorthand for a kind of unwavering commitment to what matters.

15

Running to Remember

On a crisp, starlit evening, I watch Ku Stevens run.

What you need to know about Ku Stevens running: posture relaxed, arms easy, coltish legs a blur, hot-pink track spikes flashing. In the first few laps of the men's 10,000-meter event, he is in the pack as it moves past, a single fleet organism with two dozen legs, circling around the University of Oregon's storied Hayward Field.

This is Eugene, a.k.a. Tracktown USA, home to running royalty that includes long-distance legend Steve Prefontaine, who ran under coach Bill Bowerman, who, in turn, cofounded Nike with alum Phil Knight. (My Lyft driver from the airport, a seventy-five-year-old retired banker and avid track fan, was born in Eugene and attended U of O; he let it be known that he'd run in several races with Prefontaine.)

The stadium has just been renovated to include brand-new, state-of-the-art track-and-field facilities; one level below the burnt-orange track, there is a second track underground, where training can be conducted and races run in inclement weather. A few months prior, Hayward Field hosted the World Athletic Championships, the first time the event had ever been held in the

United States. Now it is the opening meet of the NCAA spring outdoor track-and-field season, and Ku is a lean, leggy nineteen-year-old college freshman running with his dream team, the Oregon Ducks.

He has always had fast legs. I sit in the stands with Ku's parents, Misty and Delmar Stevens. Delmar says that when his son was a toddler, he would take Ku out in a stroller on his runs. Ku would sit with his sippy cup while Delmar pushed. It should be noted that Ku often got out of the stroller.

At four years old, Ku won his first race, a half-mile Jingle Bell Run. (Says Delmar: "He sprinted the whole way, arms pumping.") At eight, he won his first 5K. And it was in the last meet of his eighth-grade season that he broke his first five-minute mile.

Fast legs meant freedom. Fast legs meant fun. Fast legs meant many unlikely things he couldn't yet have imagined. In the fall of his senior year, as a cross-country team of one, from his tiny rural high school in Yerington, Nevada, Ku became a state champion: the fastest high school cross-country runner in all of Nevada.

The University of Oregon has seen many phenomenal runners. But Oregon has never had a runner like Ku. He's still just a kid, but the fact of his running contains lifetimes. His understanding of what it means to have fast legs ranges far from the track.

Hanging on the wall of his college dorm room is the pair of orange-and-white track shoes he wore when he broke the Nevada state record for the 3,200 meters in 2022. There are inscriptions on each Nike swoosh of the shoes.

On one, it reads FOR MY PEOPLE. On the other, the name of his great-grandfather: FRANK "TOGO" QUINN.

THE FIRST TIME he ran away from the Stewart Indian School, Frank Quinn was only eight years old.

The boarding school was situated at the edge of Carson City, Nevada, some fifty miles of open desert from his home on the Yerington Paiute reservation, where he was taken from his parents. The Stewart Indian School was one of more than four hundred Native American boarding schools run by the federal government in the nineteenth and twentieth centuries, operating under a policy of forced assimilation.

What this meant in practice was that young children like Frank were removed from their families against their will. They were punished for speaking their own language and forced to cut their hair and convert to Christianity. Abuse was widespread at the boarding schools. Many children died; cemeteries were found at many of the sites. Similar institutions existed in Canada. In 2021, the discovery of the remains of more than two hundred children on the grounds of the Kamloops Indian Residential School in British Columbia was a painful reminder of the trauma that continues to reverberate across Native communities, generation after generation.

As a little boy, Frank Quinn resisted. In the 1910s, he escaped Stewart and ran the fifty miles back to his family, with memory as his only guide across the treacherous arid landscape. He wanted to go home. His legs carried him there.

Federal agents showed up to bring him back to the school. He would escape and run the route twice more before school authorities gave up and left him in peace.

Ku Stevens grew up a quarter mile from where Frank Quinn lived out the rest of his days as a rancher, alfalfa farmer, miner, and respected tribal elder. As a teenage track and cross-country phenom, Ku trained in the same high-desert hills where his great-grandfather had run for his life.

Delmar Stevens remembers his grandfather as a quiet, gentle man who never spoke about what he'd endured at the Stewart Indian School. The story of his escape—once, twice, thrice, his resolve to return to his family a powerful homing beacon—was one that Delmar learned about from his mom. When Ku turned eight, the same age that Frank was when he ran for home, Delmar told him the story.

"He was real little," Delmar tells me as he watches his now-college-age son work his way around the track. "But he was the age my grandpa was when he ran away. Three times. That's what I told him—that his grandpa had the heart and determination to remain free, to exist, to not accept the genocide of who we are as human beings. We never held anything back from Ku, as Natives. He's been immersed his whole life in Paiute tradition and culture, even before he was born."

In the spring of 2021, Ku told his parents he wanted to run across the country. "He's very competitive," Misty tells me with a laugh. "He knew that there was an eighteen-year-old who had done it in a certain amount of time, and he thought that if he paced in a certain way, he could set the record for someone under eighteen. I said, 'Well, how much Gatorade is that?' We'd need a mobile home, and a crew. We didn't have the money for that. So we said, 'Let's think smaller.' Then it became, 'What about running across Nevada?' And then it was, 'What about Togo's run from Stewart to home?'" *Togo* is the Paiute word for "maternal grandfather"; Misty explains that it was also the honorific by which everyone in the community referred to Frank, whether or not he was directly related to them.

The Stevenses drove over to Carson City to visit Stewart—"to see it in the context of what it really was," Delmar explains. By then, Stewart had become a museum, but it was closed when they got there, and night was falling.

The three of them were bouncing around on a dirt road in the dark, trying to discern what it would be like to run on it, when Misty began reading on her phone. It was then that the news was breaking about the horrors unearthed at the residential school in Kamloops.

"We looked at each other," Misty says. "And then we said, 'Okay, we're doing this.'"

That summer, Ku ran the fifty-mile route home from Stewart, to honor his great-grandfather and the many children who had survived the boarding schools—and to remember those who did not make it home. More than a hundred others ran or walked with him, and Misty and Delmar pulled together the logistics to support them all, with volunteers and coordinators with the Stewart Indian Cultural Center and Museum. Misty acted as sweeper, driving the last car to make sure no one was left behind. Ku spoke to the group before they set out into the desert. "This is not a protest," he told them. "This is a remembrance." They called it the Remembrance Run.

"When I run," Ku told the *New York Times*, "I take my history with me, and especially Frank Quinn."

A run like this, tracing over his great-grandfather's footsteps, was a deliberate etching in the landscape. The day after the track meet, when Ku and I meet for breakfast at his favorite off-campus café, we talk about how it was also like the tracing of a scar. At the time of injury, the pain is sharp, acute, consuming. You can't imagine it ever getting better. Over time, the experience dulls to an ache. It becomes a memory, of something that can be healing and hopeful. In asking others to run with him, Ku hoped that they could see it with their eyes, feel it with their own bodies; it could become something new. The run was remembrance, yes, but also renewal.

"I've had my best races when I've really thought about why I'm doing it," he tells me. "I run with my heart. It's physical, but it's spiritual too. I run for my people, for their strength and resilience. Each step is a prayer. I try to remember that."

Like Pheidippides, Ku's running carries a message.

AROUND THE SAME time I watch Ku run on the track, I visit a friend elsewhere in Oregon, in the wooded parklands above Portland. I set out for a trail run among the mossy giants and try not to fall on my face as I scamper along the snow-packed trails.

Running is something that has never come easy to me, though there have been times in which I could glimpse a momentary ease, the feeling of flow. I've always preferred swimming, in which the water can absorb the work—the heat and the sweat of it—with less impact on my body. As a child, when it came to running, I had one gear: all out. What I remember most is the burning lungs and legs, feet pounding the pavement in pursuit of my father and my brother as they took off down our suburban street for the postdinner family run. Even the mile, run every year in gym class as part of the Presidential Physical Fitness Test, was like that: a hot-faced gut buster, in which I all but fell down at the finish.

I still actively dread the running portion of a triathlon; the fact that it comes last, after swimming and cycling, doesn't help the cause. These days I tend to run as a last resort when no other exercise is available to me, the ACL repair in my left knee complaining reliably after a few miles on the road, the tight IT band outside my right quadriceps piping up to say its own piece soon after.

Here in the woods, though, there are things to notice. The dull thud of a woodpecker, banging against the soft wet trunk of a tree blanketed and muted with moss. The sun-dappled path as it wends through the woods.

As I move through the forest, I have a focused awareness of my body. The dirt trail is easier on my knees. I remember what the Scottish poet and mountaineer Nan Shepherd wrote about "the long rhythm of motion sustained," that moving through a landscape is the most heightened way of experiencing your body: "The body is not made negligible, but paramount. Flesh is not annihilated but fulfilled. One is not bodiless, but essential body." The undulating path, threaded with roots, enforces a slower pace, which means I can run longer. I'm tuned in to the pulse of my heart, the swing of my arms, the feeling of my feet crunching in the snow. This awareness is entwined with the way I see the trees' outstretched arms reaching for the sky. Here, I feel there is a reason to run. The crunch I hear, the footsteps I leave—they are a temporary record, an echo of my exertions on the earth.

To physically walk another's path is more profound than you might realize. Physical movement helps you learn and remember things better—there's plenty of scientific literature that shows this correlation in the interplay between brain and body. Motor action and feedback can directly influence how well that information is absorbed and processed.

The cognitive scientist Sian Beilock has written about how the body is an important part of the learning process: Research with fMRI scans shows that "moving your foot and understanding the word 'kick' are governed by similar areas of the motor cortex," she explained. "It's hard to separate the reading mind from the doing one."

Writing by hand helps you learn in a way that's different from tapping on a keyboard—the physical movement of your hand, the pacing and tracing of the words themselves to make them your own, has much to do with committing ideas to memory.

And exercise specifically boosts neurotransmitters like dopamine, serotonin, and noradrenaline; right away, it improves our ability to pay attention. Structures of the brain involved in verbal memory and learning, like the hippocampus, get bigger with aerobic exercise. A large share of your brain's real estate is given over to movement, and new research shows that there is a connection between the effort we put into something and the reward we get from it.

We learn best when we're moving, says Anne Mangen, a professor of literacy who studies embodied cognition at the National Center for Reading Education and Research in Norway, because "our bodies are designed to interact with the world which surrounds us."

The engagement of the body in movement imprints on our brains in a distinct way. It also shakes loose new connections, new possibilities. The writer Tommy Orange once told me that embodying story while moving was something he came to rely upon: "Solutions come while running, from a part I don't access otherwise."

I read about a man, Lawin Mohammad, who is in the process of building the first long-distance hiking trail in Kurdistan. I'm struck by the way he talks about trail finding as a full-body act: "I hike this route and that route, and if I feel pain in my muscles, it's not right. Sometimes I can feel it in my teeth." A route through a landscape has a flow, a story. A good trail will take you to pretty views, pacing the ups and downs. Or sometimes the trail isn't planned for pleasure, but for something else. "It comes down to a question of why the route is here," Mohammad explains. "Why here? Why this way? Your muscles can tell why a trail exists."

A difficult route through the desert, over mountains, and among rough terrain and wildlife tells people a story that they won't easily forget, not least because they ran the path and made the journey together. Look carefully. In this shady cut is a stream; over there, a cluster of buckberries. An eight-year-old boy came this way, escaping harm. If it feels hard, it's because it *was* hard.

16

True Grit

When I consider tests of endurance, physical and otherwise, words like *outlast* and *resist* and *fight* eventually push their way to the surface. Muscle has a pugilistic quality, the capacity for violence.

Here's a true story. Sometimes, when walking alone in New York, my father got in trouble. There were attempted muggings, the odd menacing aggressor approaching on the street flinging racist words; my father answered these threats with his fists. There were times my mom packed us up in the car to pick him up late at night from the train station in our Long Island town, when he emerged from the shadows with bloodied knuckles and a story to tell.

I used to shelve these memories under Things My Crazy Dad Did, a collection that ended somewhere in the late 1990s. Now I'm not so sure where they belong. Lately, I've found myself going over the stances, the defensive blocking, the techniques of jabbing eyes and groin. I think about these things when I myself am heading out for a late night, walking alone in New York or San Francisco.

I wonder if my father trained himself and us as rigorously as he did because he thought that at some point our bodies might have to do the talking for us.

Muscle calls to mind bodyguards of dense build, towering heavies who stand sentinel at meetings between Mafia dons in the movies. Necks as thick as tree trunks, shoulders as wide as bookshelves: This is what bodies that are paid to put themselves in front of other bodies look like. Body men. Maybe you think of the Secret Service, a more streamlined presentation of muscle—the sunglasses, the suits, the closely shorn hair—drafted in the service of protecting the safety and integrity of Very Important People.

Before immigrating to New York, my father went to the same Hong Kong high school as Bruce Lee. Lee was a child star in Hong Kong before he became a sensation in the United States with a series of martial arts films that culminated in the 1973 film *Enter the Dragon*. Like Lee, my father trained in multiple disciplines of martial arts; once controversial, this mixing of training styles is now considered a precursor to the MMA fighting of today.

That kind of muscle never felt relevant to me, because muscles were never something I thought I'd have to employ in my own self-defense. Despite our father's desire to make little ninjas out of us, my brother and I were, and remain, pretty peaceable in nature. When white kids at school saw our Asian faces and asked if we knew karate, we would say, "A little? You should talk to our dad."

But we loved watching the movies of Bruce Lee, and the 1980s- and '90s-era Hong Kong martial arts films that ran on loop in our house, ones that featured stupefying stunts by action stars like Jackie Chan, Sammo Hung, and Michelle Yeoh. Our favorite parts were the outtakes at the end, where the hyperkinetic jumps, drops, and fight sequences we'd marveled at earlier were revealed to involve decidedly real bodies that regularly got broken in the attempt at perfection. Again and again, we rewound the tapes. There, in the margins of error, the pounding taken by our flesh-and-blood heroes was so palpable that our own bodies cringed and twinged on their behalf.

MAYBE I'M LYING to myself. If I think hard enough, I remember times I've flared big on the street, too—at the gang of teenage girls who shoved me on the sidewalk late one night, assuming I was a shrinking non-English-speaking immigrant; at the businessman who leaned in and whispered a demeaning sexual comment in my ear as he walked past in broad daylight, thinking his expensive suit would insulate him; at the group of men outside a bar who thought it funny to holler "Me love you long time!" at my friend and me. In moments like these, I shout back. My heart pounds, tiny muscles pull up the hairs on my neck, and I grow larger than I am, feeling the kind of fury that leads to a fight. In moments like these, I don't think about what might happen.

In contrast, I think of Kareem Abdul-Jabbar, and how I once saw him loping along at LAX, during a sleepy early-morning layover from some faraway place. The airport terminal was quiet and dim, and there he was, an elegant seven-foot-two giraffe in sunglasses coming my way. He was impossible to miss, but the thing I remember most was how he radiated calm composure.

Abdul-Jabbar was a longtime student of martial arts who'd trained with Bruce Lee. He has written about his friendship with Lee, and about how the physical and spiritual discipline in the practice Lee taught him was the foundation upon which he played two competitive decades in the NBA with little injury.

He remembers the abiding grace with which Lee handled those who were aggressive to him in public. Many times, the two would be hanging out together when someone would approach and loudly challenge Lee to a fight. "He always politely declined and moved on," Abdul-Jabbar wrote. "First rule of Bruce's fight club was don't fight—unless there is no other option. He felt no need to prove himself. He knew who he was and that the real fight wasn't on the mat."

The last few years have left many Americans shocked at the precariousness of belonging. Anti-Asian hate crimes across the country have surged: murders, acid attacks and assaults on elders, public harassment. "Get yourself some pepper spray," a friend tells me. And: "They're giving out personal alarms in Chinatown." My mother cautions me not to stand too close to the edge of the train platform. "Keep your back to the wall," she says.

To actually fight, of course, is different from being prepared to fight. Abdul-Jabbar describes quietude in the face of someone who intends to physically provoke or harm. Not using the muscle you have trained to use requires as much control as it does to weaponize it. Maybe it requires even more. To prevail over your adversary demands discipline. It asks you to hold still to your own sense of self, beyond another's breaking point.

Even as I chase the meaning of what it is to endure, I believe its highest form has to do with collective perseverance. To run with purpose can demonstrate resistance, without violence. This is how you outlast those who would hurt you. To struggle through together also changes the nature of the suffering. It makes the struggle not yours, not mine, but ours.

17

Going the Distance

The next time I see Ku Stevens run, the backdrop has shifted to Nevada's desert hues, just before sunset on a scorcher of a late-summer day. It's the afternoon before the third and final Remembrance Run. Here on the Yerington Paiute reservation, the clouds cast hypnotic shadows on wide-open spaces that move from scrub-covered flats and alfalfa farms to rocky ridgelines and, at higher elevations beyond the horizon, mountain slopes blanketed in aspen and pine. After a winter of record snowfall in the Sierra Nevada, the rivers are running high and fast, and the landscape is greener than I'd expected for August. On a grassy field next to a skate park and a sheltering stand of fruit trees, dozens of runners are setting up their tents for the night. The sky is a restlessly shifting painting, throwing light and color as its mood dictates.

I'd hitched a ride from Reno to Yerington with Paige Bethmann, a thirty-year-old Haudenosaunee filmmaker who grew up in upstate New York. We'd arrived to find Misty, Delmar, and Ku setting up shade tents and chairs for the evening meal; they greeted us as family, with big hugs. When I'd first tried to map directions from Yerington to the Stewart Indian School, Google repeatedly coughed up an error message: no route found. (Apparently, this wasn't unusual; according to Misty, Google didn't understand the reservation address or offer directions to her own house.)

Paige, though, had gotten to know these roads well. Almost two years ago, she'd moved from Brooklyn to Reno to make a documentary about Ku and his journey. Here was a young person running fifty miles across the desert to honor his ancestors. She wanted a visual record—to show the missing history, yes, but also to show someone who was remembering in order to go forward with hope.

Paige emphasizes that the Native legacy Ku carries is firmly situated in modern life. There is a long Indigenous tradition of running as a means to convey messages, prayers, and ceremonial objects; as a ritual of healing, health, and longevity; to commune with a living landscape and move through it with respect and awareness. But it isn't stuck in the past. Running is the throughline from past to present.

"Running and community are medicine," she says.

As the sun makes its way toward the horizon, I find Ku standing beneath a peach tree and looking up, thoughtful. He points out a loop of dirt roads and high trails all around us, a mental map he traces with his feet most every day when he's at home.

Ku is short for *Kutoven*, a Paiute word that means "bringing light from darkness." The social role that Ku holds in Indian country is great, and it can feel heavy to carry it all, especially for a teenager who has just finished his freshman year in college. But what I learn from spending time around Ku is that while he is contemplative and composed, he's also silly and goofy. He likes to horse around with Paige, whom he treats like a big sister. He's artistic. He designs his own tattoos and sneakers. He loves rap music.

When he runs, he doesn't really get tired. (This he says with a smile, just this side of cocky.) In high school, he didn't race anything over 5K. Because he grew up in such a small community, he'd often had to act as his own coach. Looking back at the tail end of high school, he thinks he was probably overtraining ("Running eight miles a day, at a 6:10 pace, all around these hills and trails"). But this is what he has managed to build for himself: endurance, strength, discipline. Despite the fact that he had never run distances approaching that of a marathon, much less two marathons back-to-back, this dedicated practice was what made him think he could run Frank's route home.

As young as he is, he sees how we're all connected. "It's why I'm so confident in my ability to make a change, right now," he tells me. "I *know* that I will. I *have* been. I'm able to understand how people work and what will really get through to them. A run gets their attention. And that's how I can start the conversation."

Over the last two years, the Remembrance Run has evolved into a three-act play. "My grandpa ran three times, and Ku will run three times," Delmar explains. The first run was to bring awareness. Out there, bouncing around on the dirt road, they thought: *We're going to do this, and this is why.*

The media attention and public support that followed led to the second run, which was about accountability. That event, in 2022, was run in the reverse direction, beginning in Yerington and ending at Stewart; the ceremony at the boarding school and cemetery was attended by local tribal leaders, US senator Catherine Cortez Masto, Nevada governor Steve Sisolak, and Billy Mills, the

Olympic running legend and member of the Oglala Lakota tribe who won 10,000-meter gold at the 1964 Tokyo Olympics. Mills's nonprofit youth foundation, Running Strong, funded Ku's efforts with a $10,000 Dreamstarter grant.

Delmar says that when he himself was a young man attending college in Sacramento, he was angry with the world. But he had a close friend who introduced him to the traditions of his people: ghost dance, sun dance, sweat lodge. "My friend found the Native path again to healing, and helped me to find it too," Delmar tells me. "He said, 'Culture cures, Del. Here we are still.' That's what this last run is about."

That evening, we all sit side by side at a long table, making tobacco ties to bring to the children buried in the cemetery at Stewart. "As you do this, think of a hope or prayer you might have, and put that into each one," Delmar says as he demonstrates: Take a pinch of tobacco, place it into a square of red or orange cloth, tie it off into a bundle with string.

As I begin to bundle my prayers, Pete and Stephanie Casillas, longtime friends of the Stevens family, tell stories and bestow nicknames on people in the group, as is their habit. Uncle Pete has a gray ponytail and the rumbly gravel voice of a radio announcer. When he hears that I'm a writer, he makes it a point to tell me about the first time he met Ku. "He was about knee-high, and Misty and Delmar had come to visit me in Reno," he says. The little boy hadn't yet come inside, so Uncle Pete went out the front door to find Ku, his brow furrowed with intensity, skipping back and forth over a ditch. "Like a jumping bean," Uncle Pete recalls with a laugh. "The energy and focus of those jumps? That's how I see him, even now."

Stephanie, an affectionate, round-faced woman with glasses

who dispenses hugs like candy, tells us her nickname is Froggy. "Since I'm from the Water clan," she says sweetly. "Pete is Bear, for the Bear clan. So together we're Froggy Bear." That weekend, FROGGY BEAR is scrawled merrily on their truck windows.

They christen me Scribbles; Kylie, in training to be a nurse, is given the name Scrubs; Steph, the film editor for Paige's documentary, keeps walking into things—tree branches, chair legs—so she is dubbed Dodge. With each naming, good-natured laughter echoes through the tent.

The evening wears on, and crickets and cool breezes follow. I sit there next to Ku and Paige, all of us knotting off red and orange strings of tobacco ties, and listen to Aunty Stephanie talk about her own time as a student at the Stewart Indian School. I watch other participants chat or sit quietly, lost in thought. All the while, their fingers are working: pinching, holding, tying. Passing materials to one another to share.

People talk about their families; I think especially of my sick elders, and tie prayers in their name. And I think about the ritual weaving of story and friendship at this table. It is tangible, this bond. We can feel it in our hands; over the next two days, we'd feel it with every part of our bodies.

THE NEXT MORNING, I decide to run.

This was not the plan. (Have I told you I'm not a runner?) The plan was to ride along, listen, observe, maybe walk a leg or two to talk with people as they run. But after the previous night, I want to know what it feels like to run in community—to be an active part of this collective body running fifty miles across the Nevada desert to deliver prayers to Stewart.

What surprises me about the people who show up for the

Remembrance Run is that most of them aren't really runners either. Of course, there is the group of elite high school athletes who train with Ku's former coach, Lupe Cabada, along with Cabada himself, a lean, enthusiastic man with a ready grin and a perpetual sunglasses tan. And there's an ultramarathoner from California. But there are also the Klamath River water protectors from Oregon, the musician from New Mexico, the army veteran from Utah, the reporter from New Hampshire.

Few of them will run the entire fifty-mile journey across the desert. Most are ordinary people who want to run or walk as much as they can, for deeply personal reasons, and the event is organized in a way to make this possible: twenty-five miles each day, with support vehicles and a rest stop every five miles for shade, water, and food. No one will get left behind.

As the sun angles gold across the grass, we circle up under the trees for the opening ceremony. Uncle Pete goes around to each of us and performs a smudging ritual, gently wafting smoke around our bodies with a feather as a protective blessing. Ku encourages everyone to put themselves a little bit out of their comfort zone, whether it be talking to new people or physically going the extra mile. One by one, we cross the field and assemble in the starting area. Then, with a collective whoop and cheer, we're off.

The first group of fast runners take off at a brisk clip, Ku leading the way. There is a giddiness in the air, and I find myself laughing as I run alongside my new friend Krista Fletcher, a military veteran with a wide smile and an upright bearing who had driven with a friend all the way from Boise, Idaho, in order to join this year's run.

In 2019, Krista suffered a brain injury, and lost three years of memory. "I've been putting pieces of myself back together, and finding new pieces to put in," she tells me as we turn off the paved road and head up into the hills. Part of her rehab was learning more about her Native grandfather, who'd followed the girl who would become his wife to an Indian boarding school, because he was afraid of losing her. Krista tracked down records that had been long missing from her family. Recently, she'd met with the chief of the Peoria Tribe of Indians of Oklahoma, to find out more about her grandparents' time in the boarding school. She showed me the two feathers she was running with, given to her by the tribal leader—a material connection, an anchor, to remind her that she wasn't alone.

It was another piece to add to the puzzle of who she is, but also to the puzzle of who we are as a society. "If we don't do this," she says, "how will anyone else know or remember?"

Krista had spent just shy of twenty-two years in military service: Iraq, Afghanistan, smaller tours in Cambodia. As we run, the movement jogs something free. "Part of this run, all this trauma from my military service . . ." Here, her voice trails off for a few breaths as her thoughts take shape. "I'm trying not to put the lid back on the emotion box when I start to feel."

There is something liberating about recognizing from the get-go that it's okay, and welcome, to feel things on this run, including pain and discomfort. As the sun rises higher in the sky, the pale desert terrain seems to expand before us; we talk more about feelings, and what it might have been like for an eight-year-old to be running for his life through this terrain—not once, but three times.

"Running opens up so many things for me," Krista says wonderingly. "We're running, and I'm feeling so much that I have goose bumps." She points to her legs, which are impressively prickled despite the ninety-five-degree heat. I tell her what I'd learned about goose bumps: that we have tiny muscles tied around bouquets of follicles; that they are the muscles of cold, fear, and awe; that maybe her body was allowing her to feel things again. That sometimes the body knows more than we can consciously comprehend.

With her head injury, she says, she's come to realize that she had no choice but to listen to her body: "I have to let it lead."

BY MILE EIGHT, my right quad is twanging like a too-tight drum. The blistering midday sun, the radiating heat waves off the earth, the dusty grit that infiltrates every nook and cranny. It's so dry that the sweat evaporates the moment it surfaces on my skin. Despite drinking fluids nonstop—and tossing down tiny bottles of pickle juice, courtesy of Krista, because the acetic acid in vinegar helps jolt the body out of muscle cramping—I have yet to pee this entire day.

The landscape of this part of Nevada mesmerizes: the harsh whiteout light, the ribbon of road unspooling out through the rugged terrain ahead, the sky above a cloudless and brilliant blue. I am running alone now. As I watch my shadow run in front of me—*left-right-left-right-left-right-left-right*—I listen to the way my breath heaves an echoing accompaniment of that same rhythm. *IN-hale-EX-hale-IN-hale-EX-hale-IN-hale-EX-hale.*

Sunburn, dehydration, muscle cramps, stiffness, hunger, cold: each is a bodily flash of understanding, a glimpse of Frank Quinn.

Attempting to cover more mileage in a weekend than I'd run in the previous year combined now seems painfully absurd—emphasis on the painful. My eyes sting with sweat, and a rock has worked its way into my shoe. I slow to a walk, shaking out my legs, and think about what I am here to find out. Moving through this unforgiving environment, feeling things with my body, hoping for something new to emerge from the struggle—well, we humans were built for struggle, even though most of us don't have to physically struggle in the same way to meet our needs in the modern world. What does it mean to live now, in these bodies that evolved in such a different time? Where does pain and endurance fit into the human experience at this point in our evolution? What can the body teach us, when we inch closer to its limits?

As I approach the ten-mile rest stop, my hip flexor begins to burn. Then I see the lineup of cheering figures underneath the white shade tents, Ku, Delmar, Aunty Stephanie, and Uncle Pete among them, waiting to high-five me on the way in. All else falls away, and suddenly it's clear that a smile, a kind word, and a cold drink are all I've ever wanted in life.

After a quick recharge, I ride ahead with the film crew to help set up the third rest station. By the time everyone has come in from the third leg, many runners are sitting in the shade with cramps and early signs of heat exhaustion, dousing themselves with water.

Some people turn to endurance running as a balm for grief, to create order from chaos. But there's also a physiological reward for not giving up, what psychologist Kelly McGonigal calls "the persistence high." There's no true objective measure of performance, distance, pace, she has written; instead, the reward comes from sticking with it.

That night, twenty-five miles into the run at Sunrise Pass, more than seven thousand feet above sea level, we set up camp under the pines. In the deepening twilight, far from cell service but close to the Milky Way, more than fifty of us circle up to share stories about why we are running.

"I run for my great-grandfather," Ku begins. "I'm good at

running, so that's how I honor him. Every year for the last three years, this run grounds me. It returns me to myself. It reminds me how to be open and vulnerable."

People talk about what it is to truly be in this vast landscape that Frank traversed, loving his family and his community so much he ran home to look for them.

Denise runs to pay respects to her grandparents. Jess runs to feel the struggle of the route ingrained on her body. Lupe runs for health, and for the people he loves. Krista tells everyone about her goose bumps; Ashton talks about how a little bit of physical suffering can bring spiritual and emotional understanding. "You work so hard for a sliver of peace," he says. People talk about hard things. What they did earlier in the day renders them, like Ku, unguarded and honest in this circle, grateful for this community that has come together to remember Frank Quinn, to face trauma, whether in relation to their own Native history or to other struggles.

You can't change the past. But the desire to take this experience and look forward with kindness is strong. There is a lot of beauty and joy here, despite the fact that all of our bodies are hurting. Through that effort, we come to a new understanding of each other.

At one point, I look around and think, *This is the best of humanity, right here.* This is what it means to use your body to learn, to listen, to try to feel another person's experience and story. What it means to be a runner.

Physical effort still means something in human society. To suffer physically for someone, to use your body to show that you've tried hard, has value. It's an essential way to demonstrate commitment, to earn the exchange. Think about human-powered fundraisers that generate actual dollars through movement. You walk the walk. You swim, you run, you bike—for a cause, a cure, a community. It shows you have skin in the game. As illogical as it may seem, it is a way to demonstrate care.

That night in circle, Paige talks about what it means that, for the first time in three years of filming, she is finally getting to run herself.

"It was so hard," she tells me later. "I am not really a runner. But I really needed that firsthand perspective." She could feel the heat through her shoes; the land felt endless. She was grateful she had Ku's sister Star running by her side. The only way she could keep moving was to look at the clouds, to find a sense of hope that she would make it. "I was proud that I was able to finish five miles," Paige says. "But *fifty* miles? How do you do that? To be a little kid, to want to make it home. Connection, community, and choosing that over anything. I thought about kids crossing the border. My fingers were swelling, my legs were chafing, but I would have been disappointed in myself if I didn't try. If I didn't suffer."

THE NEXT DAY will bring a hypnotic wave of heat, a blur of aches and pains. Near the end of the run, there will be a river crossing outside Carson City. We'll plunge into the cool water, finding relief before floating back up to the vivid blue sky. Later, I'll wonder: *Did Frank find relief in that creek?* The last two hundred yards before reaching the Stewart Indian School, we will gather and walk together into the cemetery, led by Ku, with Billy Mills at his side. We will be greeted by young dancers in full regalia, who are now the age of the children in the cemetery when they died. We will place our prayers and our tobacco ties on those tiny headstones. There will be tears of emotion and exhaustion. After, the dancers will do the joyful bodily thing that helps us to commit this day to memory. They will fill our spent hearts.

The writer Rachel Kushner once observed that "to endure or tolerate or last are all terms that indicate a person withstands some kind of burden to which they have been subjected." To embark on a long and difficult walk in which one focuses on the journey rather than the destination, say, is to *choose* to be subjected to that burden, because the granularity of the experience alters one's understanding of time—one might say to the tune of enlightenment: "Human steps dilate the integers of time."

How we interpret time passing depends on how it's kept—by a second hand, a beating heart. Sixty to a hundred beats per minute, a hundred thousand beats a day, thirty-five million beats a year. Three billion beats or more in an average lifetime. The heart is its own clock, muscle arranged in bundles to move blood in the right direction. To trace a human stride provides yet another dimension to the understanding of movement, and of time; it is the holding of a collective present and past. Your path, and someone else's.

For the Nobel Prize–winning muscle scientist Albert Szent-Györgyi, muscle as a lens to look at the world made perfect sense. He discovered the actin and myosin proteins, and how the addition of ATP results in the fundamental reaction of muscle contraction. In describing the brilliant scientist, the biologist Steven Vogel wrote that, for Szent-Györgyi, "movement defined animal life, muscle enabled movement"—think *animate*—and that by studying muscle, "he could come closest to the essence of life."

In our animal bodies, endurance taken too far is destructive.

Muscle ceases to be the engine and, instead, becomes fuel as the body consumes itself. Endurance for its own sake knows no end. But if humans revere the push to the edge, perhaps it's because we've found that endurance for a purpose—pushing through, to reach a new state of equilibrium, in order to achieve that purpose—alchemizes it into something else entirely.

18

Kissing the Ground in Equilibrium

Let us return, at long last, to the humble push-up.
Every day, it's a simple test: you versus you. Get down, face the floor, hold yourself on hands and feet, lower your body down as a single straight plank, feel the tension in wrists, arms, shoulders, and back, lift up. Repeat. Lift and lower, lift and lower. Some days you feel it more in your neck; on others, maybe it's your back or butt or hamstrings that speak up first. Most physical scientists agree that the push-up is the enduring barometer of fitness. It's whole-body movement with a functional purpose.

The push-up materializes in morning boot camp with a friend, in exercises I do with the boys, in pop-ups on my surfboard. You're down, then you get back up, again and again. (If that isn't a suitable metaphor for life, I don't know what is.) It has a particular utility, as a daily measure of how you carry yourself. Over time, that adds up.

"It takes strength to do them, and it takes endurance to do a lot of them," said the fitness pioneer Jack LaLanne, who might have known this best. He became synonymous with push-ups when he got Americans to exercise with him on *The Jack LaLanne Show*, which aired on television for more than thirty years, from 1951 to 1985. LaLanne once set a world record by doing a thousand push-ups in twenty-three minutes.

Jack LaLanne's meet-cute story with his wife, Elaine, actually involves push-ups: It was 1951, on the San Francisco set of the *Les Malloy Show*, and Elaine Doyle was the talent booker for the televised variety program, which had a twelve-piece band and often featured movie stars who came to town. She'd received an intriguing call about a guy in Oakland, California, who could do push-ups throughout the entire ninety-minute show. *Oh boy, that sounds like fun,* thought Elaine. *We'll put him in the corner and let him do his thing—just pan over to him every once in a while.* Somewhere between the first push-up and the one-thousandth—or maybe it was the five hundredth one after that—Jack looked up and saw Elaine. As of this writing, Elaine LaLanne is still doing push-ups in California, at the age of ninety-seven.

It turns out that even Jan Todd, the strength athlete and sports historian, has something to learn from this foundational movement. Recently, someone sent a request to the Stark Center about the origin of the push-up, and it reignited a memory that she'd long forgotten, of an illustration of a small girl doing modified push-ups by placing her hands on her father's knees; it was in a book by the nineteenth-century exercise expert Phokion Heinrich Clias.

Jan knew that the push-up likely wouldn't have been called a *push-up* before a certain time in history, and she was right. After some time in the archives, she unearthed the English-language copy of Clias's 1819 book, *An Elementary Course of Gymnastic Exercises*, first published in French, in which she found the exercise, back when Clias called it "Kissing the Ground in Equilibrium on the Arms and the Points of the Feet." It still makes her smile when she says it.

In her findings, Jan says that the point of her inquiry wasn't actually to discover the origin of the push-up:

"There will never be a way to say definitively who did the first push-up, or the first squat, or the first handstand. Maybe it was a Spartan warrior, or an early Indian wrestler, or perhaps a knight of the Middle Ages who needed more strength to wield his broadsword. Someone was undoubtedly the first to fall forward to 'kiss the ground' so they could build more strength in their upper body, but we will never know [their] name."

What we learn in the process, Jan says, is that for the past two hundred years or so, humans haven't been just doing push-ups; we've really been kissing the ground in equilibrium, through all the ups and downs—and it makes her feel stronger to know it.

LIKE THE PUSH-UP, the marathon is a test. Endurance sports are proxies for sticking it out, struggle, strength. They can be powerful forms of communication.

Society keeps erasing and forgetting Indigenous communities. Ku Stevens's run is a way to remember and to spread the word.

"A lot of people from other tribes, farther away, have come to join us," Misty Stevens tells me. "They're on their own journeys. And a lot of the non-Indian community is really supportive, the people who come out. It's good to hear their words too; they weren't taught this in school. They come on the journey to learn about it."

An elementary school teacher from Markleeville, California, named Jennifer Munyan tells me she first attended the

Remembrance Run after finding out about it from the on-campus Native representative at her school. Back at home, she took what she'd learned about Indian boarding schools and integrated it into the fifth- and sixth-grade history curriculum, and is planning to lead a trip to Stewart with many of her students.

"I ran thirty miles last year, and I thought of my students a lot when I was out there in the heat," Jennifer says, about the way that physical effort leaves an impression. "My body suffered immensely the second day, and my mind returned to the outrageousness of an eight-year-old being so distraught that his best hope was to run fifty miles across a desert." She made a prayer necklace of tobacco ties for each of her students, and thought about their families as she ran.

It was upon arrival at the cemetery, after enduring the heat that summer day, that she was overwhelmed with emotions. "There were last names I knew—Christensen and Nevers, last names of Washoe friends, of families of the students I serve, real stories of people I know," she tells me. "There's a dark side to humanity, and it's our obligation to help our students understand what atrocities come from ignorance and hate." A powerful answer to generational trauma is lasting generational change.

Not long ago, Ku appeared on the cover of a youth magazine that was distributed to schools all across the country. Misty thinks it was likely the first time that the history of Indian boarding schools was presented so widely in the US public school system. "That was a gigantic victory," she says. "A tangible way that the run made a difference. A whole generation is being exposed to this information. And that's all because of Ku."

For a Native teenager who has spent his young life watching Native history be erased, this is no small thing. Breaking records, being first—it's the most obvious way of making sure you will be remembered. But it's not the only way.

"A lot of these guys on my team, they're insanely talented," Ku tells me. "And me? You know, I'm *pretty* talented. But a big part of what I run for and how I run so well: I run with my heart. In all races, you hit a point: 'Am I gonna go? Am I gonna quit?' It's a choice between one or the other, going or quitting. And those are the moments when you could really use some help. A reminder of why you're doing this, to catch a second wind."

So much of endurance is having a good reason to persist, when everything in your body is telling you to stop.

"In that moment? I usually just say 'For my people.' Just those three words," he says. "And then I go."

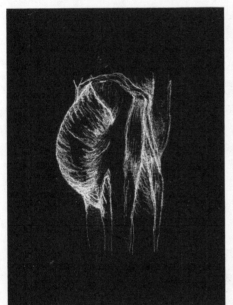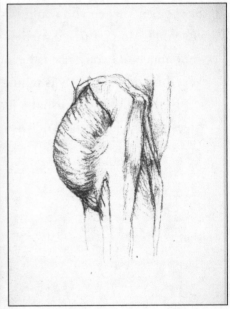

Gluteus maximus.
Illustration by the author

EPILOGUE

One morning, on the phone in the dark as I drive to the surf, my father asks me what surfing does for me: "What's it like to walk on water?"

What I tell him about surfing is that it feels like magic. Despite the expenditure of energy, surfing and swimming return me to my body in the best ways. It's restoring a version of myself that can handle hard things with grace.

He tells me he, too, rises in the predawn hours, to run around the neighborhood before anyone else is up—in the dark, like we used to. He gets up early so he can move through the world before it wakes and fills with other people. "You know how I hate crowds," he says. He knows that in the afternoons there are grandmothers who chatter by the playground bar where he does his pull-ups and flips and other body-weight exercises. Were they to spot him working out, they'd wave and shout "Hey! You there! What's that you're doing?" He doesn't feel like explaining himself. So he gets up early, just like I do, to be in the world in the quiet hour. The time of solitude, filled with muted air and light. The timing of our practice, in the early hours, is no coincidence.

Sometimes I think I'm muscling it too much, trying too

hard—with surfing, with my dad. And then I remember the movement and the joy—how muscles move us toward happiness. The poet Sharon Olds, told not to cross her legs after a bout of hip surgery, could barely contain her legs' playfulness. She hadn't realized how vain she was about her legs until she was supposed to keep them still, she wrote—"no / semaphoric waving, no / Rockette Rockette-flanked."

The most joyful thing I've been practicing is walking on water. When I do, I try not to run, because running on a surfboard is a recipe for disaster. (No semaphoric waving.) Unlike land, water is an unstable platform; on a longboard, measured, unhurried steps, with control, is the goal.

Which brings us to grace. Anton Chekhov described grace as occurring "when someone expends the least amount of motion on a given action." Grace, in part, is about restraint, physical calm in the face of uncertainty.

There was a time when the physics of surfing seemed impossible to me. What was involved in order to dance on the water with grace was mind-boggling: spotting a wave, negotiating a constantly moving and changing surface, paddling into the right position, getting to your feet, finding balance, keeping your feet connected to the board *and also* moving those feet to shift your weight, all while reading the changes in the water and maintaining speed. Not to mention having an awareness of countless other bodies and boards that are also moving.

Proprioception is a four-dimensional ability—not just the awareness of how your body is moving in three planes, but the predictive ability over time to know where you will be to meet the next wave.

There's no way to get around paying your dues in the water—to learn all the many and specific variables, and then to be capable of synthesizing it all into a reasonable gestalt for various situations. The symphony of muscles acting together takes practice. There's a temporal aspect, because it's something you repeat to the tune of many years. Grace, then, is also about endurance.

It's about endurance in the nearer term too. When I emerge from a long surf, run, or swim, the tiredness in my muscles tells me that I've done something worthwhile. The state of aching is not always a bad one; in this case, the residual burn is a symptom of what it means to really live a life of the body. The restless vibration of my spine has quieted; my body is relaxed, and so is my mind.

"Our bodies prime our metaphors," observed the writer James Geary, "and our metaphors prime how we think and act." Muscle is metaphor and memory.

The effect of this muscle exploration is that I've had to participate in my own studies of character. Strength, form, action, flexibility, endurance: All of these are qualities of muscle that we strive for in personhood. The constant telescoping between the physical and the philosophical is revealing. With grace, the beauty and elegance in movement comes from a seeming effortlessness and ease. But we all know that nothing comes without hard work. Let us look instead at the effort behind the apparent ease. Let us honor all the time and work—the stretching, heaving, grimacing—that has gone into achieving the oneness of your body with the world, the connection you can feel with something, or someone, outside of yourself. To be in the ballet of life around you. What I take away from the enterprise is the complexity of all that muscle signifies about what we desire and who we want to be.

IN THE SUMMER of 2023, I convince my father to meet us in Hong Kong.

When we are together in person, I can't help but marinate in the joy. We walk in step together, arms touching. We talk exercise routines. He squeezes my shoulder. We visit the Hong Kong Museum of Art. As we roam around the galleries, he trains his artist eye on me. The first thing he comments on is my strength: "You look good. You look strong." It's a different spin on how most people might greet each other, but it makes sense to me, because this is how my dad looks at the world. I'm happy to see that he looks strong too. He asks Matt to judge the width of our shoulders as we stand back-to-back, in front of a painting of a swimmer.

He stops to examine a large, framed piece of art with Teddy. Their contemplative stances are visual echoes of each other, the way they incline their heads, the way they hold their hands. Teddy, now ten, has a keen talent for art, and his work is full of personality. He paints portraits of friends, characters from favorite books and movies, California poppies for his grandmother, hermit crabs encountered on a weekend at the beach. He is perceptive with his seeing, and the way he listens to everyone in a room. His touch is delicate.

In the museum, my father talks to him about painting—watercolor, gouache, oil—and about how it's okay to make mistakes. He invites Teddy to apprentice in his studio sometime. Teddy listens carefully and nods, pleased in the glow of his grandfather's attention. They agree to become painting pen pals, exchanging art across the Pacific.

"You send me one, and I'll send you one," my father says to Teddy. "Okay?"

Teddy nods his assent, then looks at me. "Okay, Mama?"

We come across a hands-on exhibit on calligraphy overlooking the harbor. Printed on the floor-to-ceiling glass windows are Chinese characters, for *water* and *sky* and *sea* and *wind*. The exhibit invites visitors to practice calligraphy with water on a quick-drying surface with three different kinds of brushes; after a few minutes, the writing will dry and disappear, so that someone else can try their hand at it.

We each make our choice of tool, and, without even thinking, my father and I immediately start doing quick brush sketches of each other. There's a giddiness to this paint-off: He paints me with my hair up in a bun; I paint him with his baseball cap and his laughing open mouth. Felix and Teddy giggle as they watch, and Matt steps back to take a photo of us all.

"This is why you can't paint family," my father observes sagely. "Because they always think you got them wrong."

Maybe I got him wrong. Or maybe what I have learned through this inquiry into muscle is to allow him the room to stretch, to grow, to change. I can't make my father be someone he isn't. But muscle holds the idea of capability. What is it but potential, the promise of future action?

He says to me in the museum, "So what do you want to ask me?"

I laugh in response, because on some level he knows I have been waiting to receive some kind of message from him. And I say, simply, honestly: "I just want to be with you."

BACK AT HOME, Teddy and I draw together. I'm reminded that visual art is a record of movement—observing a moving subject, capturing with a roving hand, revealing to an appreciating eye. Art is different for me now, but I like the notion that it is once again a new thing—a language to speak with my son. In the creation of the muscle illustrations that begin and end this book, he has been good company. In between drawing sessions, we'll do a set of burpees or take a few laps around the block, a relic of pandemic PE that has stuck around in our household. We are developing our very own version of Muscle Academy.

The other day, Teddy painted his self-portrait, and instructed me to send it to my dad for feedback. My father praised the distinct character that Teddy had managed to capture that made him who he was.

Nature and nurture aren't really opposed: In every family, there's a perpetual relationship between them. We try to please our parents, and we try to please our children. In a self-repeating pattern, we will forever be trying to please them both. Somewhere in the negotiation between the two, there's a chance for rediscovery.

Make me a muscle. I understand that doing so was the first time I felt capable. This fundamental pride in self, propelling me through life.

I'm reminded of a rare visit my father made years ago, when the boys were small, to see us in California. As he flipped through a sketchbook I'd started with them, he'd encouraged me to draw seriously again. "Why not?" he'd asked. With that, he made me feel something that I hadn't in a long time, and it wasn't about the art.

He made me feel that I could do anything.

ACKNOWLEDGMENTS

Soon enough, this book will get up and walk around on its own. Before it does, I want to thank the many generous and wonderful humans who helped me bring it to life.

My deepest gratitude to the incredible Jan Todd; Dan O'Conor, Ebony Ingram, Joy Jones, and Robbin Ebb, who prove that everyday levitation is possible; my father-in-law, Jim Elliott, who shared Jumbo, and himself; Matthew Sanford, Angelique Lele, and Molly Bachman, who know what grace is; and the Stevenses—Ku, Misty, and Delmar—who welcomed a stranger into their extended family and told her to stay awhile. I'm forever changed by all of you. Thank you to Paige Bethmann and the She Carries Her House crew for taking care of me in the desert. I can't wait to see what comes.

Huge thanks to Amber Fitzsimmons, Maddie Norris, Barbie Klein, and Dana Rohde, who revealed the magic of the body as so much more than the sum of its parts, and to Daniel Wolpert, Dacher Keltner, Adam Sharples, Daniel Turner, Magne Lund-Hansen, Harris Masket, and Kevyn Dean, who patiently walked me through the science of all things muscle and movement. This book would not have been possible without your kindness and expertise. Special thanks to UCSF, the Norwegian School of Sport Sciences, and the British Library for the gift of access to your worlds of knowledge.

To Danielle Svetcov, agent, superhero, dear friend. What can I say except it just keeps getting better and better thanks to you.

Big love to Lynsay Skiba, Jenny Fu, Caroline Paul, Rachel Levin, and Chris Colin: trusted readers, cheerleaders, the scaffolding for my brain and heart.

To my brilliant editor, Amy Gash: I knew from the first you were the one for me. Algonquin Books really is my family and I love you all. Debra Linn, Michael McKenzie, Brunson Hoole, Betsy Gleick, Christopher Moisan, Cathy Schott, Steve Godwin . . . thank you for seeing my vision and being on my team. Steve Kalinda, thank you for your beautiful cover art. Elizabeth Johnson, you're a copy-editing genius and a true gem.

I can do what I do because of 360-degree wraparound love and support from so many: Anna Vella, Esther Chak, Mara Gladstone, Frances Duncan, James Wilson, Tom Davidson, Melissa Gibson, Michele Caputo, Mirek Boruta. You mean so much to me. Thank you, Sarah McCarthy, for boot camps and belly laughs. To Zouhair Belkoura, for dawn patrols and digressions. To Jessica Bath, for holding my hand with kindness and dressing me properly for the anatomy lab.

Thanks to the Mesa Refuge, a serendipitous place that has long been special to my writing life. It's also where I realized that I really did want to write this book (Liz Rosner, your grace and good vibes got my brain and broken ribs back to where that seemed possible again!). Huge love to Rebecca Skloot, for inviting me to the purple dream house. To Lynda and Bob Balzan, always, for sharing that Bolinas magic.

Books take forever. I'm so grateful to every editor who let me try out the bits, pieces, and weird little experiments that eventually got me here, and to the writing communities—the Writers Grotto and the Notto among them—that have helped sustain me throughout.

To my mom, the steady foundation of love upon which I'm built.

To my dad, my perennial sounding board for all things art and exercise.

To my brother, Andy, my first and best sparring buddy and true friend.

And the biggest gratitude and love go to my three favorite people: Matt, Felix, and Teddy. You make the impossible possible, even pandemic P.E. Anything and everything is worth doing when it's with you.

NOTES

Introduction

3   *The biologist and biomechanics pioneer Steven Vogel wrote*   Steven Vogel, *Prime Mover: A Natural History of Muscle* (New York: W. W. Norton, 2003).

3   *recently discovered fossil of a cnidarian*   Lucas Leclère et al., "Diversity of Cnidarian Muscles: Function, Anatomy, Development and Regeneration," *Frontiers in Cell and Developmental Biology*, January 23, 2017.

3   *first flexing their muscles*   University of Cambridge, "Animals First Flex Their Muscles: Earliest Fossil Evidence for Animals with Muscles," *ScienceDaily*, August 26, 2014, https://www.sciencedaily.com/releases/2014/08/140826205417.htm.

3   *Strongest and biggest*   "What Is the Strongest Muscle in the Human Body?" *Everyday Mysteries: Fun Science Facts from the Library of Congress*, Library of Congress, https://www.loc.gov/everyday-mysteries/biology-and-human-anatomy/item/what-is-the-strongest-muscle-in-the-human-body.

4   *brain health depends on muscles*   Alinny R. Isaac et al., "How Does the Skeletal Muscle Communicate with the Brain in Health and Disease?" *Neuropharmacology* 197 (October 1, 2021).

5   *Paul Klee described visual art as a record of movement*   "The pictorial work was born of movement, is itself recorded movement, and is assimilated through movement (eye muscles)," wrote Klee in his 1920 critical text, "Creative Confession."

STRENGTH

7   *Muscles are in a most intimate and peculiar sense the organs of the will*   G. Stanley Hall, *Youth: Its Education, Regimen, and Hygiene* (New York: D. Appleton, 1906).

What's Power, in a Body?

9   *In a 1990 interview*   Gary Groth, "Jack Kirby Interview," *The Comics Journal* #134 (February 1990), https://www.tcj.com/jack-kirby-interview/6.

10   *what has come to be known as hysterical strength*   Carly Vandergriendt, "How Superhuman Strength Happens," Healthline, July 23, 2020, https://www.healthline.com/health/hysterical-strength; Jessica Firger, "The Science Behind Superhero Strength," CBS News, July 2, 2014, https://www.cbsnews.com/news/the-science-behind-superhero-strength; Jeff Wise, "When Fear Makes Us Superhuman," *Scientific American*, December 28, 2009, https://www.scientificamerican.com/article/extreme-fear-superhuman.

10   *When Jan Todd was a girl*   I spoke with Jan Todd personally about her experiences, and her story has been widely reported in magazines and newspapers, including Sarah Pileggi, "The Pleasure of Being the World's Strongest Woman," *Sports Illustrated*, November 14, 1977, https://vault.si.com/vault/1977/11/14/the-pleasure-of-being-the-worlds-strongest-woman.

11   *traditions that extend as far back as 2000 BCE*   "About Highland Games," Scotland.org, https://www.scotland.org/events/highland-games/about-highland-games.

11   *heavy lifting stones, or clachan togail, were sometimes used to prove manhood*   Peter J. Martin researched and wrote extensively about Gaelic strength and stone lifting; his books and articles are compiled on the websites https://www.oldmanofthestones.com and https://www.thedinniestones.com.

11   *Todd became the first woman ever to lift the Dinnie Stones*   Jan Todd and others talk about historic feats of strength in the Scottish Highlands in the 2016 documentary *Stoneland*, available for viewing at https://www.youtube.com/watch?v=MhQlNwxn5oo.

12   *The British neuroscientist and Columbia University professor Daniel Wolpert calls himself a "movement chauvinist."*   I spoke with him in person about muscles and mind, and he gave a 2011 TED talk on the topic called "The Real Reason for Brains," https://www.ted.com/talks/daniel_wolpert_the_real_reason_for_brains.

13   *the arctic tern, perhaps the ultimate endurance athlete*   "To the Ends of Earth," a *National Geographic* article on the annual migration of the arctic tern, https://education.nationalgeographic.org/resource/ends-earth.

13   *male white-spotted pufferfish, who spends a week or more*   "The Pufferfish," from episode 1 of the 2017 PBS documentary *Big Pacific*, https://www.pbs.org/video/pufferfish-f7eual.

13   *Male fiddler crabs have one larger claw*   Sophie L. Mowles et al., "Multimodal Communication in Courting Fiddler Crabs Reveals Male Performance Capacities," *Royal Society Open Science* 4 (March 2017).

14   *Birds of prey*   "Extraordinary Female Animals," BBC Earth, https://www.bbcearth.com/news/extraordinary-female-animals.

14   *scientists theorize that their larger size*   Kate Fallon, "Most Female Raptors Are Bigger and Stronger Than Males, but Why?" Audubon, March 12, 2018, https://www.audubon.org/news/most-female-raptors-are-bigger-and-stronger-males-why.

14   *men have 80 percent more muscle mass*   David Epstein, *The Sports Gene: Inside the Science of Extraordinary Athletic Performance* (New York: Penguin, 2013).

16   *Todd would later write*   Terry Todd, *Inside Powerlifting* (Chicago: Contemporary Books, 1978).

18   *coached the Texas Longhorns*   Jan Todd showed me a commemorative T-shirt from the time that listed all the dates of championship trophies.

18   *Scientists have identified a number of genes that are involved in muscular development and growth*   Sara M. Willems et al., "Large-Scale GWAS Identifies Multiple Loci for Hand Grip Strength Providing Biological Insights into Muscular Fitness," *Nature Communications* 8 (2017); Sander A. J. Verbrugge et al., "Genes Whose Gain or Loss-Of-Function Increases Skeletal Muscle Mass in Mice: A Systematic Literature Review," *Frontiers in Physiology* 9 (2018). Transgenic mouse models have helped identify genes whose gain or loss of function results in muscle hypertrophy in mice; in the prominent case of MSTN, a loss of MSTN in mice *or* humans roughly doubles muscle mass.

19   *Recent research on testosterone-induced muscle growth*   Oscar Horwath et al., "Fiber Type-Specific Hypertrophy and Increased Capillarization in Skeletal Muscle Following Testosterone Administration in Young Women," *Journal of Applied Physiology* 128 (May 2020).

20   *she would later write, about besting her father*   Jan Todd, "Father to a Strongwoman," *Aethlon*, Spring 2002.

Muscle as Potential

22 *murals depicting the exercise can be found in the tombs of Beni Hasan*   Marley Brown, "Emblems for the Afterlife," *Archaeology*, May/June 2018, https://www.archaeology.org/issues/296-1805/features/6508-egypt-middle-kingdom-tomb-paintings; "Fundamental Principles of Olympism," Egyptian National Olympic Committee, https://www.egyptianolympic.org/Olympism.html.

22 *They had ritual significance*   Elizabeth Childs-Johnson, "Big Ding and China Power: Divine Authority and Legitimacy," *Asian Perspectives* 51 (Fall 2012); Sima Qian, trans. Burton Watson, *Records of the Grand Historian, Qin Dynasty* (New York: Columbia University Press, 1993). I also spoke with my father regarding Chinese-language sources about the ding, Sima Qian, and King Wu of Qin.

23 *"burdensome stone"*   William Smith, *A Concise Dictionary of the Bible* (London: W. Clowes and Sons, 1865); I also read various translations and commentaries gathered at Bible Hub, https://www.biblehub.com/commentaries/zechariah/12-3.htm.

23 *In Japan, thousands of chikairashi . . . In Icelandic culture*   I consulted Japanese and Icelandic colleagues on the lore around these stone-lifting cultures, as well as research articles, including Sepp Linhart, review of *Japanese Sports: A History*, *The Journal of Japanese Studies* 29 (2003); John Marshall Carter, *Ritual and Record: Sports Records and Quantification in Pre-Modern Societies* (Westport, CT: Greenwood, 1990).

24 *you can find the Naha Stone on the Big Island*   Spencer Kealamakia, "Remember the Naha Stone," *Hawai'i Magazine*, November 7, 2017, https://www.hawaiimagazine.com/remember-the-naha-stone.

25 *muscles that "jump out at you"*   Barry McDermott, "Trying to Muscle In," *Sports Illustrated*, October 21, 1974, https://vault.si.com/vault/1974/10/21/trying-to-muscle-in.

25 *nicknamed the Iron Paradise*   Phil Blechman, "25 Years of Transformation—The Physique of Dwayne 'The Rock' Johnson," BarBend, July 13, 2023, https://www.barbend.com/dwayne-the-rock-johnson-physique-transformation.

27 *what sportswriters have called "the great separator"*   Pileggi, "The Pleasure of Being the World's Strongest Woman."

28 *"The highest competitive two-handed lift by a woman is 392 lbs."*   Pileggi, "The Pleasure."

29 "*if the strength being considered is muscle strength*" Pileggi.

29 "*I love the way it makes me feel*" Pileggi.

29 *Jan appeared on The Tonight Show* Clip available at http://www.youtube.com/watch?v=t1QG49Csc7w.

A Heavy Lift

31 "*In all my reading nothing seemed quite so wonderful*" Terry Todd, "A Legend in the Making," *Sports Illustrated*, November 5, 1979.

31 *At the 1889 World's Fair in Paris* Joshua Robinson, "When It Comes to Scottish Games, Americans Are Plaid to the Bone," *New York Times*, July 24, 2007, https://www.nytimes.com/2007/07/24/sports/othersports/24highland.html. Jan Todd also shared writings on the Dinnie Stones by the historian David Webster.

32 *a passage that seemed to be a direct invitation to Jan* David Webster, *Scottish Highland Games* (Edinburgh: Reprographia, 1973).

32 *women were first allowed to compete in the 1900 Olympics* "FAQ: History and Origin of the Games," International Olympic Committee, https://www.olympics.com/ioc/faq/history-and-origin-of-the-games/when-did-women-first-compete-in-the-olympic-games; "Factsheet: The Games of the Olympiad," IOC, October 2013, accessed at the Wayback Machine, http://www.web.archive.org/web/20200422134610/https://stillmed.olympic.org/Documents/Reference_documents_Factsheets/The_Olympic_Summer_Games.pdf.

32 *The two granite stones together weigh 733 pounds* For many years, the Dinnie Stones were thought to weigh a total of 788 pounds, as described in Terry Todd's November 1979 *Sports Illustrated* story. In 2014, Jan Todd explained, the stones were reweighed and the results certified.

35 "*The two brutal-looking rocks were chained together*" Todd, "A Legend in the Making."

37 *where a full stadium tour is a leg-extinguishing thirty-five hundred steps* Rounded up from 3,441 steps. Elizabeth Martin explains the math of a full Harvard Stadium tour workout in "An Ode to Harvard Stadium," www.gentlegiant.com/blog/ode-harvard-stadium.

38 *Weightlifting has been shown to build resilience in the mind* Danielle Friedman, "The Healing Power of Strength Training," *New York Times*, July 7, 2022, https://www.nytimes.com/2022/07/07/well/move/weight-lifting-ptsd-trauma.html.

The Making of a Hero, Then and Now

40 *"Joe, we need muscles like yours to beat Germany"* William Dettloff, "The Louis-Schmeling Fights: Prelude to War," HBO Boxing, accessed at the Wayback Machine, http://www.web.archive.org/web/20090529151439/http://www.hbo.com:80/boxing/features/history/joe_louis.html.

40 *he had gained about that many pounds in muscle* Burkhard Bilger, "The Strongest Man in the World," New Yorker, July 16, 2012, https://www.newyorker.com/magazine/2012/07/23/the-strongest-man-in-the-world.

41 *a documentary about Terry* More information on the film, The Commissioner of Power, can be found at https://starkcenter.org/2021/10/rogue-fitness-honors-terry-todd-with-documentary-premiering-at-austin-film-festival.

42 *"the history of men's sports is uninterrupted mythmaking"* Kate Fagan, "What Would Happen if Women Athletes Got the Mythology Treatment They Deserve?" (guest essay), New York Times, April 3, 2023, https://www.nytimes.com/2023/04/03/opinion/women-sports.html.

43 *a story about Pa O'Dwyer* You can read Malachy Clerkin's original profile, "Pa O'Dwyer Is Ireland's Strongest Man—the Six Steaks a Day Way," published in the Irish Times, December 1, 2018, https://www.irishtimes.com/sport/pa-o-dwyer-is-ireland-s-strongest-man-the-six-steaks-a-day-way-1.3715960.

45 *"Even if somebody is born with a particular talent"* Yuval Noah Harari, Sapiens: A Brief History of Humankind (New York: HarperCollins, 2015).

46 *Jan was codirecting the 2022 Arnold Strongman Classic* The Arnold Strongman Classic is part of the Arnold Sports Festival, which bills itself as the world's biggest multi-sport fitness festival.

FORM

49 *"Who has not been, or is not a good master of the figure"* In 1560, Michelangelo wrote this in a letter to Cardinal Ridolfo Pio of Carpi. J. M. S. Pearce, "The Anatomy of Michelangelo (1475–1564)," Hektoen International: A Journal of Medical Humanities 11 (Spring 2019), https://hekint.org/2018/04/11/anatomy-michelangelo-1475-1564.

The Ideal Body

51   *a state of contraction called rigor mortis*   "Rigor Mortis," ScienceDirect, https://www.sciencedirect.com/topics/medicine-and-dentistry/rigor-mortis.

52   *the fog here has a name*   Karl the Fog can be found at https://twitter.com/karlthefog and https://www.instagram.com/karlthefog/

53   *"Chinese doctors lacked even a specific word for 'muscle'"*   Shigehisa Kuriyama, *The Expressiveness of the Body and the Divergence of Greek and Chinese Medicine* (New York: Zone Books, 2002).

54   *The ideal body in ancient China*   Shigehisa Kuriyama discusses at length the historical contrast between Western and Eastern medical conceptions of the body and its ideals in *The Expressiveness of the Body and the Divergence of Greek and Chinese Medicine*. This section quotes from Kuriyama's chapter on muscularity and identity.

55   *the Vatican's completed restoration of Michelangelo's Sistine Chapel ceiling*   The restoration ran from 1980 to 1994 and occurred in stages: the lunettes (completed in 1984), the ceiling (completed in 1989), *The Last Judgment* (completed in 1994). Susanne Simpson and Peter Thomas, "Saving the Sistine Chapel," a television documentary of *Nova* (WGBH/Boston, with NTV International, 1989), available at https://www.archive.org/details/CantheVaticanSavetheSistineChapel.

55   *"see the blood flowing through their veins"*   Meg Nottingham Walsh, "Out of the Darkness: Michelangelo's Last Judgment," *National Geographic*, May 1994.

55   *Michelangelo's intimate knowledge of human anatomy*   Pearce, "The Anatomy of Michelangelo."

56   *highly restricted by the Catholic Church*   Alison Abbott, "Hidden Treasures: Padua's Anatomy Theatre," *Nature* 454 (2008), https://www.nature.com/articles/454699a.

56   *Artists couldn't easily obtain bodies to dissect*   "Leonardo and Dissection," from the catalogue for the exhibition *Leonardo da Vinci: Anatomist: The Queen's Gallery, Buckingham Palace*, Royal Collection Trust, https://www.rct.uk/collection/themes/exhibitions/leonardo-da-vinci/the-queens-gallery-buckingham-palace/leonardo-and-dissection.

56   *elevated the portrayal of the human body*   Sanjib Kumar Ghosh, "Human Cadaveric Dissection: A Historical Account from Ancient Greece to the Modern Era," *Anatomy & Cell Biology* 48 (September 2015).

56   *"a tumultuous sea of jumbled bodies"*   Walsh, "Out of the Darkness."

56   *a satirical sonnet*   Michelangelo, *The Complete Poems of Michelangelo*, trans. John Frederick Nims (Chicago: University of Chicago Press, 1998). The full poem can be found at https://press.uchicago.edu/Misc/Chicago/080331.html.

58   *"You have farmed with your uncle"*   These lines appeared as marketing copy for Choose Your Own Adventure #109. R. A. Montgomery, *Chinese Dragons* (New York: Bantam Books, 1991).

59   *a volcanic expression of the soul*   Francis Henry Taylor, "Michelangelo: The Titan and the Crisis," *The Atlantic*, September 1955, https://www.theatlantic.com/magazine/archive/1955/09/michelangelo-the-titan-and-the-crisis/640968.

Who's Afraid of a Lady Hercules?

61   *muscular women have "disrupt[ed] the equation"*   David L. Chapman with Patricia Vertinsky, *Venus with Biceps: A Pictorial History of Athletic Women* (Vancouver, BC: Arsenal Pulp, 2011).

61   *a period of nearly two centuries*   Writer Maria Popova features a gallery of images from *Venus with Biceps* in her blog, *The Marginalian*, https://www.themarginalian.org/2011/11/21/venus-with-biceps.

62   *twenty-first century Saudi Arabia*   Minky Worden, "Saudi Arabia's Newest Sportswashing Strategy: Sponsorship of Women's World Cup," Human Rights Watch, February 16, 2023, https://www.hrw.org/news/2023/02/16/saudi-arabias-newest-sportswashing-strategy-sponsorship-womens-world-cup.

62   *the ballet superstar Misty Copeland*   Jill Radsken, "Misty Copeland, Offstage," *Harvard Gazette*, May 10, 2017, https://news.harvard.edu/gazette/story/2017/05/dancer-misty-copeland-shares-her-life-story-with-students.

62   *the tennis great Serena Williams*   Paula Cocozza, "Serena Williams: 'Not Everyone's Going to Like the Way I Look,'" *Guardian*, June 28, 2016, https://www.theguardian.com/lifeandstyle/2016/jun/28/serena-williams-interview-beyonce-dancing-too-masculine-too-sexy.

63   *establishment of nonbinary divisions*   Kelyn Soong, "Boston Is a Bucket-List Marathon. Now Nonbinary Runners Can Compete, Too," *Washington Post*, April 10, 2003, https://www.washingtonpost.com/wellness/2023/04/10/boston-marathon-nonbinary-runners.

63   *testosterone is the hormone that actually makes people reckless*   Gideon Nave et al., "Single-Dose Testosterone Administration Impairs Cognitive Reflection in Men," *Psychological Science* 28 (October 2017).

63 *pseudoscience has long governed norms around women's anatomy and biology* Amanda Loudin, "After Decades of Neglecting Women Athletes, Sport and Exercise Medicine Is Finally Catching Up," Stat, May 19, 2023, https://www.statnews.com/2023/05/19/sports-medicine-women-athletes.

67 *across multiple studies of NCAA athletes in different sports* Jason Shurley et al., "Historical and Social Considerations of Strength Training for Female Athletes," *Strength and Conditioning Journal* 42 (August 2020).

68 *Netter Atlas of Human Anatomy* Frank Netter, *Netter Atlas of Human Anatomy*, 8th ed. (Cambridge, MA: Elsevier, 2022).

68 *a deeply troubling vein of white supremacy* Michael Paterniti, "The Most Dangerous Beauty," *GQ*, September 28, 2002, https://www.statnews.com/2023/05/19/sports-medicine-women-athletes.

68 *the ethics of teaching from this book continue to be widely debated* Andrew Yee et al., "Ethical Considerations in the Use of Pernkopf's Atlas of Anatomy: A Surgical Case Study," *Surgery* 165 (May 2019).

68 *The social psychologist Jaclyn A. Siegel has studied* Sam Risak, "The Strong, Silent Type: Jaclyn A. Siegel on Masculinity and Male Body Image," *Sun*, March 2023, https://www.thesunmagazine.org/issues/567/the-strong-silent-type.

69 *the quiet increase of boys and men seeking help* Jason Fuller, Bridget Kelley, and Juana Summers, "Eating Disorders in Young Men Are Being Masked by Muscle Bulking and Over-Exercising," *All Things Considered*, National Public Radio, July 27, 2023, https://www.npr.org/2023/07/27/1190578569/eating-disorders-in-young-men-are-being-masked-by-muscle-bulking-and-over-exerci.

69 *A recent headline from a normally quite staid Harvard Medical School publication* The May 18, 2023, edition of the *Harvard Gazette* featured this headline, referring to the article by Maureen Salamon, "A Muscle-Building Obsession in Boys: What to Know and Do," Harvard Health Publishing, https://www.health.harvard.edu/blog/-a-muscle-building-obsession-in-boys-what-to-know-and-do-202305122934.

69 *Italian art restorers were faced with a strange problem* Jason Horowitz, "Send in the Bugs. The Michelangelos Need Cleaning," *New York Times*, May 31, 2021, https://www.nytimes.com/2021/05/30/arts/bacteria-cleaning-michelangelo-medici-restoration.html.

Shoulders, Squared

71 *"There are few body parts"* Vanessa Friedman, "How Wide Can You Go?," *New York Times*, March 30, 2023, https://www.nytimes.com/2023/03/28/style/big-shoulders-trend.html.

74 *Vesalius himself—considered the founder of modern anatomy* Fabio Zampieri et al., "Andreas Vesalius: Celebrating 500 Years of Dissecting Nature," *Global Cardiology Science & Practice* 5 (2015), https://www.ncbi.nlm.nih.gov/pmc/articles/PMC4762440.

75 *"How we use a muscle"* Vogel, *Prime Mover*.

75 *We have fast-twitch and slow-twitch muscle fibers* Jennifer Harbster, "Please Pass the Slow-Twitch Fiber," *Inside Adams* (blog), Library of Congress, November 24, 2009, https://blogs.loc.gov/inside_adams/2009/11/please-pass-the-slow-twitch-fiber%E2%80%A6; Vogel, *Prime Mover*.

76 *On average, women have a larger percentage of slow-twitch fibers* Christine Yu has a comprehensive discussion of what makes women particularly well suited physiologically to ultra-endurance activities in "The Long Game" chapter of her book, *Up to Speed: The Groundbreaking Science of Women Athletes* (New York: Riverhead, 2023).

76 *A study of identical twins* Alex Hutchinson, "Nature vs. Nurture: Study on Twins Shows Athletic Destiny Not Set at Birth," *Globe and Mail*, August 17, 2018, https://www.theglobeandmail.com/life/health-and-fitness/article-nature-vs-nurture-study-on-twins-shows-athletic-destiny-not-set-at.

76 *In a procedure called dynamic cardiomyoplasty* Mariell Jessup, "Dynamic Cardiomyoplasty: Expectations and Results," *Journal of Heart and Lung Transplantation* 19 (August 2000), https://doi.org/10.1016/S1053-2498(99)00104-7.

78 *In the early 1960s, the neurologist and writer Oliver Sacks* Oliver Sacks, *On the Move: A Life* (New York: Knopf, 2015).

78 *"I was accepted on Muscle Beach"* Sacks, *On the Move*.

78 *"My motive, I think, was not an uncommon one"* Sacks, *On the Move*.

78 *"far beyond their natural limits"* Sacks, *On the Move*.

78 *"While I was in hospital in 1984"* Sacks, *On the Move*.

86 *Skeletal muscle fibers are unique*   "Skeletal Muscle," Cleveland Clinic Health Library, https://my.clevelandclinic.org/health/body/21787-skeletal-muscle#overview; Physiopedia, "Muscle Cells (Myocyte)," Physiopedia, https://www.physio-pedia.com/Muscle_Cells_(Myocyte); "Skeletal Muscle Cell," ScienceDirect, https://www.sciencedirect.com/topics/medicine-and-dentistry/skeletal-muscle-cell; Vogel, *Prime Mover*.

ACTION

89 *"Muscles are highly expressive"*   G. B. Duchenne de Boulogne, *The Mechanism of Human Facial Expression*, trans. and ed. R. Andrew Cuthbertson (New York: Cambridge University Press, 1990).

89 *"Is smiling a practice?"*   Sarah Ruhl, *Smile: The Story of a Face* (New York: S&S/Marysue Rucci Books, 2021).

Your Muscles Are Talking

92 *On the one-year anniversary*   "'Great Lake Jumper' Dan O'Conor Has Jumped into Lake Michigan Every Day for Over a Year, and Has Come Out Better for It," CBS News, July 25, 2021, https://www.cbsnews.com/chicago/news/great-lake-jumper-dan-oconor-lake-michigan.

92 *via social media, I watch Dan throw himself into the lake*   You, too, can follow the Great Lake Jumper: https://twitter.com/TheRealDtox and https://www.instagram.com/danielt.oc.

94 *scouts have referred to a prospective player's strong gluteus maximus*   Josh Kendall, "A High 'Butt Factor' Might Be an NFL Draft Prospect's Most Prized Asset," *Athletic*, March 21, 2024, https://theathletic.com/5351145/2024/03/21/nfl-prospect-draft-butt-factor-evaluation.

94 *the best of them have jumps as good as NBA point guards*   In 2018 and 2019, I spent several months with Kevyn Dean and USA Surfing reporting on biomechanics research they were doing in the lead-up to the 2020 Summer Olympics in Tokyo; Bonnie Tsui, "Surfers Are Riding a Wave of New Technologies to Their Olympic Debut," *Popular Science*, July 10, 2020, https://www.popsci.com/story/technology/surfing-olympics-training-technology.

95 *When our muscles need energy to move*   "Mitochondria," an episode of the BBC podcast *In Our Time*, June 1, 2023, https://www.bbc.co.uk/programmes/m001md34.

95 *the neurochemical signal cascade* Alinny R. Isaac et al., "How Does the Skeletal Muscle Communicate with the Brain in Health and Disease?" *Neuropharmacology* 197 (October 2021).

96 *formation of new neurons* Gretchen Reynolds, "How Exercise Leads to Sharper Thinking and a Healthier Brain," *Washington Post*, April 5, 2023, https://www.washingtonpost.com/wellness/2023/04/05/exercise-brain-thinking-bdnf.

96 *boost learning and memory* "Cognitive Health and Older Adults," National Institute on Aging, https://www.nia.nih.gov/health/brain-health/cognitive-health-and-older-adults.

96 *your muscles and brain are talking to each other* Connie Chang, "New Clues Are Revealing Why Exercise Can Keep the Brain Healthy," *National Geographic*, June 6, 2022, https://www.nationalgeographic.com/magazine/article/new-clues-are-revealing-why-exercise-can-keep-the-brain-healthy.

96 *release of endocannabinoids and dopamine* Sandra Amatriain-Fernández et al., "The Endocannabinoid System as Modulator of Exercise Benefits in Mental Health," *Current Neuropharmacology* 19 (August 2021).

96 *"a pharmacy in your muscles"* Kelly McGonigal talks with Shawn Stevenson in "The Extraordinary Link Between Exercise, Joy, and Human Connection," episode 393 of *The Model Health Show*, January 14, 2020, https://themodelhealthshow.com/kelly-mcgonigal.

96 *Research on bed rest reveals* Alexandra Kleeman, "The Bed-Rest Hoax," *Harper's Magazine*, December 2015, https://harpers.org/archive/2015/12/the-bed-rest-hoax.

Jumpology

99 *Charles Darwin had a thing or two to say* Charles Darwin, *The Expression of the Emotions in Man and Animals* (New York: D. Appleton, 1872).

99 *brain-muscle cross talk* Darwin's Manuscripts Collection, Cambridge University Library, available through the American Museum of Natural History, https://www.amnh.org/research/darwin-manuscripts/catalogue-darwin-manuscripts/cambridge-university-library.

100 *emotions can be thought of as "relational acts between people"* Nikhil Krishnan, "How Universal Are Our Emotions?" *New Yorker*, August 8, 2022, https://www.newyorker.com/magazine/2022/08/08/how-universal-are-our-emotions.

100 *"many of our emotion terms are references to states of the body"* Krishnan, "How Universal Are Our Emotions?"

101 *lasting influence on fields* André Parent, "Duchenne de Boulogne: A Pioneer in Neurology and Medical Photography," *Canadian Journal of Neurological Sciences* 32 (2005).

101 *Studies show that jumping* Ayelet Melzer et al., "How Do We Recognize Emotion from Movement? Specific Motor Components Contribute to the Recognition of Each Emotion," *Frontiers in Psychology* 10 (2019).

102 *"collective effervescence"* Dacher Keltner introduced me to this phrase, attributed to the French sociologist Émile Durkheim.

102 *Scientists who study the physics of breaching* Paolo S. Segre et al., "Energetic and Physical Limitations on the Breaching Performance of Large Whales," *eLife*, March 11, 2020, https://doi.org/10.7554/eLife.51760.

103 *dorsal camera-tag footage of a fifty-ton humpback ascending to the surface to breach* Segre et al., "Energetic and Physical Limitations on the Breaching Performance of Large Whales." The footage of a fifty-ton humpback (described in the paper as forty-six thousand kilograms) breaching is available for viewing here.

103 *the weight of a large Gulfstream jet* A Gulfstream G650 weighs about one hundred thousand pounds, https://www.jetcraft.com/jetstream/2022/03/gulfstream-g650-overview-2012-present-2.

103 *research suggests that it's about communicating location or activity* "Scientists Finally Figure Out Why Whales Like to Jump Out of the Water," *Discover*, November 29, 2016, https://www.discovermagazine.com/planet-earth/scientists-finally-figure-out-why-whales-like-to-jump-out-of-the-water.

104 *"popping with life"* Annie Dillard, *Pilgrim at Tinker Creek* (New York: Harper Perennial Modern Classics, 2013).

106 *almost two hundred jumps a minute* James Thomas, "Training for Double Dutch at the Apollo," *New York Times*, December 4, 2021, https://www.nytimes.com/interactive/2021/12/04/nyregion/double-dutch-contest-apollo.html. Competitive jumpers average roughly 380 jumps in two minutes.

106 *Every year since 1992* From the official website of the National Double Dutch League, www.nationaldoubledutchleague.com/holiday-classic.

106 *an oral and kinetic tradition* "How the Jump Rope Got Its Rhythm," a TED talk by ethnomusicologist and TED Fellow Kyra Gaunt, http://www.ted.com/talks/kyra_gaunt_how_the_jump_rope_got_its_rhythm/transcript.

109 *in 2004, when she started the DC Retro Jumpers* Sharon McDonnell, "Spreading the Joy of Double Dutch," *AARP*, October 2021, https://www.aarp.org/home-family/friends-family/info-2021/author-creates-double-dutch-team.html.

110 *"To move things is all that mankind can do"* J. C. Eccles and W. C. Gibson, *Sherrington: His Life and Thought* (Berlin: Springer, 1979); Helge Nørstrud, ed. *Sport Aerodynamics* (Vienna: Springer, 2009).

111 *a record 101 Life magazine covers* "Philippe Halsman: Jump," Magnum Photos, http://www.magnumphotos.com/arts-culture/philippe-halsman-jump-book; Owen Edwards, "When He Said 'Jump . . .'", *Smithsonian*, October 2006, https://www.smithsonianmag.com/arts-culture/when-he-said-jump-130897523.

111 *"In a jump, the subject, in a sudden burst of energy"* "Philippe Halsman: Jump," Magnum Photos.

112 *"I realized that deep underneath people wanted to jump"* Philippe Halsman, *Philippe Halsman's Jump Book* (New York: Damiani, 2015).

112 *"Jumping humanity can be divided into two categories"* Halsman, *Philippe Halsman's Jump Book*.

112 *"The Interpretation of Jumps"* Halsman, *Jump Book*.

113 *"After all, maybe, this is not a bad way to go"* Halsman, *Jump Book*.

113 *"I hope it will mean to the judge what it means to me"* Halsman, *Jump Book*.

The Movement Is the Message

116 *Movement, Stulberg has written, "demands you pay close attention"* Brad Stulberg, "The Benefits of Daily Movement," *Outside*, September 7, 2021, https://www.outsideonline.com/health/wellness/daily-movement-benefits-practice-groundedness-stulberg.

118 *wrote about how his specific physical capabilities* Russell Janzen, "On Leaving the Life of the Body: A Dancer Reports," *New York Times*, September 20, 2023, https://www.nytimes.com/2023/09/20/arts/dance/dancer-retirement-new-york-city-ballet.html.

FLEXIBILITY

121 *"He has changed externally"* Carlos Montezuma, "Changing Is Not Vanishing," Academy of American Poets, https://poets.org/poem/changing-not-vanishing.

Muscles, Fast and Slow

123 *one of the greatest track-and-field coaches in history* Thomas Rogers, "Jumbo Elliott of Villanova Is Dead: Long an Outstanding Track Coach," *New York Times*, March 23, 1981, https://www.nytimes.com/1981/03/23/obituaries/jumbo-elliott-of-villanova-is-dead-long-an-outstanding-track-coach.html.

124 *Frank Budd ran for Jumbo* Roy Terrell, "Record Dash En Route to Moscow," *Sports Illustrated*, July 3, 1961, https://vault.si.com/vault/1961/07/03/record-dash-en-route-to-moscow; Dave D'Alessandro, "Sprinter's World-Record Legacy Lives On at the Frank Budd Track Meet in Asbury Park," *Star-Ledger*, July 6, 2013, https://www.nj.com/ledger-dalessandro/2013/07/sprinters_world-record_legacy_lives_on_at_the_annual_frank_budd_track_meet_at_asbury_park_hs.html; Frank Litsky, "Frank Budd, Once Known as World's Fastest Human, Dies at 74," *New York Times*, May 1, 2014, https://www.nytimes.com/2014/05/01/sports/frank-budd-once-known-as-worlds-fastest-human-dies-at-74.html; Gerald Holland, "Here Comes Jumbo," *Sports Illustrated*, January 22, 1962, https://vault.si.com/vault/1962/01/22/here-comes-jumbo.

124 *"I guess I just didn't notice"* D'Alessandro, "Sprinter's World-Record Legacy Lives On."

124 *a remarkably detailed "day in the life" profile of Jumbo* Holland, "Here Comes Jumbo."

126 *"His life began with polio"* D'Alessandro, "Sprinter's World-Record Legacy Lives On."

127 *Back in the fall of 1978* I spoke with Matthew Sanford personally about his experiences and read numerous magazine and newspaper stories about his work. Sanford is also the author of the memoir *Waking* (Emmaus, PA: Rodale, 2006).

129 *an ancient discipline tracing back thousands of years in India* Sat Bir Singh Khalsa, "Is Yoga 5,000 Years Old? The Archaeology of Yoga," Kundalini Research Institute, https://kundaliniresearchinstitute.org/en/is-yoga-5000-years-old-the-archaeology-of-yoga.

It Comes from Unity

141 *Muscle spindles are our secret stretch detectors* "Muscle Spindle," ScienceDirect, https://www.sciencedirect.com/topics/neuroscience/muscle-spindle.

141 *Peripersonal space representation* Michele Scandola et al., "Visuo-Motor and Interoceptive Influences on Peripersonal Space Representation Following Spinal Cord Injury," *Scientific Reports* 10 (2020).

142 *Interoception is your body's ability to sense itself from inside* Wen G. Chen et al., "The Emerging Science of Interoception: Sensing, Integrating, Interpreting, and Regulating Signals within the Self," *Trends in Neurosciences* 44 (January 2021); Jessica Wapner, "The Paradox of Listening to Our Bodies," *New Yorker*, July 6, 2023, https://www.newyorker.com/science/elements/the-paradox-of-listening-to-our-bodies.

142 *A recent study with paraplegic and control subjects* Scandola, "Visuo-Motor and Interoceptive Influences."

144 *the ribbon tied around a bunch of flowers* Niloufar Torkamani et al., "Beyond Goosebumps: Does the Arrector Pili Muscle Have a Role in Hair Loss?" *International Journal of Trichology* 3 (July–September 2014).

145 *a sign that we are trying to grasp the unknown* Laura A. Maruskin et al., "The Chills as a Psychological Construct: Content Universe, Factor Structure, Affective Composition, Elicitors, Trait Antecedents, and Consequences," *Journal of Personality and Social Psychology* 103 (2012).

145 *our muscles are twitching in REM sleep* Amanda Gefter, "What Are Dreams For?" *New Yorker*, August 31, 2023, https://www.newyorker.com/science/elements/what-are-dreams-for.

145 *The neuroscientist Mark Blumberg* Gefter, "What Are Dreams For?"

148 *Matthew once told the interviewer Krista Tippett* "Matthew Sanford: The Body's Grace," *On Being*, October 5, 2006, https://onbeing.org/programs/matthew-sanford-the-bodys-grace-2023.

148 *In the poem "A Suit or a Suitcase"* Maggie Smith, "A Suit or a Suitcase," *Nation*, November 23, 2022, https://www.thenation.com/article/culture/a-suit-or-a-suitcase.

Remembrance of Exercises Past

154 *When we talk about muscle memory* Adam P. Sharples and Daniel C. Turner, "Skeletal Muscle Memory," *American Journal of Physiology: Cell Physiology* 324 (2023).

155   *In 2018, his research group was the first in the world*   Robert A. Seaborne et al., "Human Skeletal Muscle Possesses an Epigenetic Memory of Hypertrophy," *Scientific Reports* 8 (2018).

155   *They have a lasting molecular memory*   Gretchen Reynolds, "How 'Muscle Memory' May Help Us Get in Shape," *New York Times*, January 5, 2022, https://www.nytimes.com/2022/01/05/well/move/muscle-memory-exercise.html.

155   *Cellular muscle memory, on the other hand, works a little differently*   E. J. Foulstone et al., "Adaptations of the IGF System during Malignancy: Human Skeletal Muscle versus the Systemic Environment," *Hormone and Metabolic Research* 35 (2003). A good explainer on cellular muscle memory is Stacey Colino's article "Here's What Muscle Memory Really Means, and How to Use It," *Washington Post*, August 9, 2022, https://www.washingtonpost.com/wellness/2022/08/09/muscle-memory-motor-skills-fitness.

156   *he began to investigate the why*   Adam P. Sharples et al., "Skeletal Muscle Cells Possess a 'Memory' of Acute Early Life TNF-α Exposure: Role of Epigenetic Adaptation," *Biogerontology* 17 (2016).

156   *The research also has far-reaching implications*   "Study Proves 'Muscle Memory' Exists at a DNA Level," Keele University, January 30, 2018, https://www.keele.ac.uk/research/researchnews/2018/january/studyprovesmusclememoryexistsatadnalevel/muscle-memory.php.

156   *Norway is expected to become a "super-aged society"*   AARP International's 2018 country report on Norway, sourcing the United Nations Population Division of the Department of Economic and Social Affairs, https://www.aarpinternational.org/initiatives/aging-readiness-competitiveness-arc/norway.

157   *a normal process known as sarcopenia*   R. Roubenoff, "Sarcopenia and Its Implications for the Elderly," *European Journal of Clinical Nutrition* 54 (2000); T. Priego et al., "Role of Hormones in Sarcopenia," *Vitamins and Hormones* 115 (2021); Han Shu et al., "An Integrated Study of Hormone-Related Sarcopenia for Modeling and Comparative Transcriptome in Rats," *Frontiers in Endocrinology* 14 (2023).

157   *women who do weight training just a couple of days a week*   Allison Aubrey, "Women Who Do Strength Training Live Longer. How Much Is Enough?" *Morning Edition*, National Public Radio, March 17, 2024, https://www.npr.org/2024/03/17/1239036584/women-who-do-strength-training-live-longer-how-much-is-enough.

158   *Globally, Scandinavian countries*   Dinesh K. Dhanwal et al., "Epidemiology of Hip Fracture: Worldwide Geographic Variation," *Indian Journal of Orthopaedics* 45 (2011).

158 *Falls are the second-leading cause of unintentional death worldwide* World Health Organization fact sheet on falls, April 26, 2021, https://www.who.int/news-room/fact-sheets/detail/falls.

158 *In Norway, hip fracture caused by accidental falling* Norwegian government data shared with me by Sharples and his team.

158 *In the Netherlands, there are even standardized classes on how to fall* Christopher F. Schuetze, "Afraid of Falling? For Older Adults, the Dutch Have a Cure," *New York Times*, January 2, 2018, https://www.nytimes.com/2018/01/02/world/europe/netherlands-falling-elderly.html.

159 *a woman in her seventies who changed her life with weightlifting* Katie Kindelan, "How This 75-Year-Old Woman Lost over 60 Pounds, Became a Fitness Influencer," *Good Morning America*, March 16, 2023, https://www.goodmorningamerica.com/wellness/story/75-year-woman-lost-60-pounds-fitness-influencer-82655773.

163 *the gene called UBR5 affects muscle size and recovery* Robert A. Seaborne et al., "UBR5 Is a Novel E3 Ubiquitin Ligase Involved in Skeletal Muscle Hypertrophy and Recovery from Atrophy," *Journal of Physiology* 597 (2019).

164 *serum from the blood of hibernating black bears* Mitsunori Miyazaki et al., "Supplementing Cultured Human Myotubes with Hibernating Bear Serum Results in Increased Protein Content by Modulating Akt/FOXO3a Signaling," *PLoS One* 17 (2022); Nick Lavars, "Serum from Hibernating Black Bears Boosts Muscle Mass in Human Cells," *New Atlas*, July 18, 2022, https://newatlas.com/science/serum-hibernating-black-bears-muscle-mass-human-cells.

164 *Lifelong exercisers in their seventies* Stacey Burling, "It's Never Too Late to Start Moving, but Science Is Finding You May Not Catch Up to Lifelong Exercisers," *Philadelphia Inquirer*, September 14, 2021, https://www.inquirer.com/health/lifelong-exercisers-versus-people-who-start-late-muscle-mass-strength-sedentary-20210914.html.

165 *Newly popular techniques like blood flow restriction* Matthew Futterman, "A Hot Fitness Trend among Olympians: Blood Flow Restriction," *New York Times*, July 21, 2021, https://www.nytimes.com/2021/07/21/sports/olympics/athletes-blood-flow-restriction.html.

166 *with repeated exercise, immune T cells* Ekaterina Pesheva, "Research Shows Working Out Gets Inflammation-Fighting T Cells Moving," *Harvard Gazette*, November 3, 2023, https://news.harvard.edu/gazette/story/2023/11/new-study-explains-how-exercise-reduces-chronic-inflammation.

166 *Stronger leg muscles* Jacqueline Howard, "Strong Leg Muscles May Be Linked with Better Outcomes after Heart Attack, Study Suggests,"

CNN, May 29, 2023, https://www.cnn.com/2023/05/29/health/leg-muscles-heart-attack-study-wellness/index.html.

166 *Even grip strength*   Darryl P. Leong et al., "Prognostic Value of Grip Strength: Findings from the Prospective Urban Rural Epidemiology (PURE) Study," *Lancet* 386 (2015).

166 *After five months of aerobic exercise training*   I discussed these findings with Adam Sharples; Piotr P. Gorski et al., "Aerobic Exercise Training Resets the Human Skeletal Muscle Methylome 10 Years after Breast Cancer Treatment and Survival," *FASEB Journal* 37 (January 2023), https://doi.org/10.1096/fj.202201510RR.

ENDURANCE

169 *"To stretch, heave, grimace"*   Margo Jefferson, *Constructing a Nervous System* (New York: Pantheon Books, 2022).

What We Carry

171 *back to the year 490 BCE*   Judith Swaddling, the British Museum's senior curator of Greece and Rome, wrote about the ancient Olympics and illustrated it with specific museum holdings in "The Marathon's Ancient Origins," British Museum, September 11, 2017, https://www.britishmuseum.org/blog/marathons-ancient-origins.

171 *Herodotus sang the praises*   Herodotus, "Battle of Marathon: 6.94–117," Greek and Roman Historians, the Latin Library, https://www.thelatinlibrary.com/historians/herod/herodotus8.html.

172 *"So, when Persia was dust"*   Robert Browning, "Pheidippides," from *Browning's Shorter Poems* (1899), available at https://www.gutenberg.org/files/16376/16376-h/16376-h.htm.

172 *"Joy in his blood bursting his heart"*   Browning, "Pheidippides."

172 *billions of pairs of Nikes*   It is widely reported that Nike sells more than 780 million pairs of shoes a year; Aditya Misra, "Nike Doesn't Sell Shoes. It Sells an Idea with Its Marketing Strategy," The Strategy Story, July 30, 2020, https://www.thestrategystory.com/2020/07/30/nike-marketing-strategy.

173 *Adaptations that aided in this pursuit*   Daniel E. Lieberman et al., "Running in Tarahumara (Rarámuri) Culture: Persistence Hunting, Footracing, Dancing, Work, and the Fallacy of the Athletic Savage," *Current Anthropology* 61 (2020); Luke A. Kelly et al., "Active Regulation of Longitudinal Arch Compression and Recoil during Walking and Running," *Journal of the Royal Society: Interface* 12 (2015).

173 *the human body is able to respond dramatically to physical demands* Christopher McDougall, *Born to Run: The Hidden Tribe, the Ultra-Runners, and the Greatest Race the World Has Never Seen* (New York: Knopf, 2009).

173 *Blood flow to skeletal muscles can increase up to a hundred times* Mayo Clinic faculty profile of Michael J. Joyner, whose study areas include blood flow during exercise and the physiology of elite athletes, https://www.mayo.edu/research/faculty/joyner-michael-j-m-d/bio-00078027.

174 *As science writer Heather Radke puts it* Heather Radke, *Butts: A Backstory* (New York: Avid Reader Press, 2022).

174 *Daniel Lieberman explains, running is really a kind of controlled falling* William J. Cromie, "Running Paced Human Evolution: Anthropologists Conclude Running May Have Helped Build a Bigger Brain," *Harvard Gazette*, November 18, 2004, https://www.news.harvard.edu/gazette/story/2004/11/running-paced-human-evolution.

174 *"a multifunction Swiss Army knife"* Radke, *Butts: A Backstory*. Radke interviewed University of Colorado Boulder scientist Jamie Bartlett, who called the gluteus maximus "a multifunction Swiss Army knife."

174 *"the will to endure can't be reliably tied"* Alex Hutchinson, *Endure: Mind, Body, and the Curiously Elastic Limits of Human Performance* (New York: Mariner Books, 2018).

174 *In the late stages of prolonged exercise* Suzanne Girard Eberle, *Endurance Sports Nutrition* (Champaign, IL: Human Kinetics, 2014).

175 *a recent paper about humans and long-distance running* Lieberman, "Running in Tarahumara (Rarámuri) Culture."

Running to Remember

177 *In the fall of his senior year* Kurt Streeter, "To Honor His Indigenous Ancestors, He Became a Champion," *New York Times*, November 17, 2021, https://www.nytimes.com/2021/11/17/sports/ku-stevens-running-nevada.html.

177 *he broke the Nevada state record for the 3,200 meters* Chris Murray, "Yerington's Ku Stevens Wins Second Nevada Gatorade State Athlete of the Year Award," Nevada Sports Net, June 29, 2022, https://www.nevadasportsnet.com/news/reporters/yeringtons-ku-stevens-wins-second-nevada-gatorade-state-athlete-of-year-award; Bill Donahue, "Running to Remember: Ku Stevens," Red Bull Bulletin, March 8, 2023, https://www.redbull.com/us-en/theredbulletin/ku-stevens-long-distance-running-native-american-remembrance-run.

178   *the discovery of the remains of more than two hundred children*   Ian Austen, "'Horrible History': Mass Graves of Indigenous Children Reported in Canada," *New York Times*, May 28, 2021, https://www.nytimes.com/2021/05/28/world/canada/kamloops-mass-grave-residential-schools.html.

180   *"I take my history with me"*   Streeter, "To Honor His Indigenous Ancestors, He Became a Champion."

182   *"the long rhythm of motion sustained"*   Nan Shepherd, *The Living Mountain: A Celebration of the Cairngorm Mountains of Scotland* (Edinburgh: Canongate Books, 2008).

182   *Research with fMRI scans shows*   Sian Beilock, "How Humans Learn: Lessons from the Sea Squirt," *Psychology Today*, July 11, 2012, https://www.psychologytoday.com/us/blog/choke/201207/how-humans-learn-lessons-the-sea-squirt.

183   *Writing by hand helps you learn in a way that's different*   University of Stavanger, "Better Learning through Handwriting," *ScienceDaily*, January 24, 2011, https://www.sciencedaily.com/releases/2011/01/110119095458.htm.

183   *new research shows that there is a connection*   Kelly Lambert, a professor of behavioral neuroscience at the University of Richmond, studies effort-based rewards, and found that hands-on activities and physical effort support learning; Markham Heid, "Working with Your Hands Is Good for Your Brain," *New York Times*, March 28, 2024, https://www.nytimes.com/2024/03/28/well/mind/hands-mindfulness-typing-writing.html.

183   *"our bodies are designed to interact with the world which surrounds us"*   University of Stavanger, "Better Learning through Handwriting."

183   *I read about a man, Lawin Mohammad*   Ben Mauk, "Building the First Long-Distance Hiking Trail in Kurdistan," *New York Times Magazine*, April 24, 2022, https://www.nytimes.com/2022/04/20/magazine/hiking-kurdistan.html.

True Grit

188   *He has written about his friendship with Lee*   Kareem Abdul-Jabbar, "Bruce Lee Was My Friend, and Tarantino's Movie Disrespects Him," *Hollywood Reporter*, August 16, 2019, https://www.hollywoodreporter.com/tv/tv-news/kareem-abdul-jabbar-bruce-lee-was-my-friend-tarantinos-movie-disrespects-him-1232544.

188   *"He always politely declined and moved on"*   Abdul-Jabbar, "Bruce Lee Was My Friend."

Going the Distance

191 *to make a documentary about Ku and his journey* Paige Bethmann's film is called *Remaining Native*, https://www.remainingnativedocumentary.com.

192 *the ceremony at the boarding school* Jim Krajewski, "Visiting Children's Graves in School Cemetery Gives Added Meaning to Remembrance Run," *Reno Gazette Journal*, August 16, 2022, https://www.rgj.com/story/sports/2022/08/16/remembrance-run-ku-stevens-reno-nevada-stewart-indian-school/10334078002.

197 *acetic acid in vinegar helps jolt the body out of muscle cramping* Stephanie E. Hooper Marosek et al., "Quantitative Analysis of the Acetic Acid Content in Substances Used by Athletes for the Possible Prevention and Alleviation of Exercise-Associated Muscle Cramps," *Journal of Strength Conditioning Research* 34 (June 2020).

198 *"the persistence high"* Kelly McGonigal, *The Joy of Movement: How Exercise Helps Us Find Happiness, Hope, Connection, and Courage* (New York: Avery, 2021).

201 *"to endure or tolerate or last"* Rachel Kushner, "Learning to Wait," *Harper's Magazine*, October 2023, https://harpers.org/archive/2023/10/learning-to-wait.

201 *"Human steps dilate the integers of time"* Kushner, "Learning to Wait."

201 *"movement defined animal life"* Vogel, *Prime Mover*.

Kissing the Ground in Equilibrium

203 *"It takes strength to do them, and it takes endurance"* Tara Parker-Pope, "An Enduring Measure of Fitness," *New York Times*, March 11, 2008, https://www.nytimes.com/2008/03/11/health/nutrition/11well.html.

204 *Jack LaLanne's meet-cute story with his wife, Elaine* "The Push-Up: A Fitting Fitness Test," National Public Radio, March 14, 2008, https://www.npr.org/2008/03/14/88236377/the-push-up-a-fitting-fitness-test; Donald Katz, "Jack LaLanne Is Still an Animal," *Outside*, November 1995, https://www.outsideonline.com/outdoor-adventure/jack-lalanne-still-animal; Danielle Friedman, "At 97, the First Lady of Fitness Is Still Shaping the Industry," *New York Times*, October 3, 2023, https://www.nytimes.com/2023/09/04/well/move/elaine-lalanne-fitness-exercise.html.

204 *an illustration of a small girl doing modified push-ups* Jan Todd, "An Unexpected Pleasure: Kissing the Ground in Equilibrium—A Search for

the Origins of the Pushup," The H. L. Lutcher Stark Center for Physical Culture and Sports, https://starkcenter.org/2023/01/an-unexpected-pleasure-kissing-the-ground-in-equilibrium-a-search-for-the-origins-of-the-pushup.

204 *"Kissing the Ground in Equilibrium"* Peter Henry Clias, *An Elementary Course of Gymnastic Exercises; Intended to Develope and Improve the Physical Powers of Man* (London: Sherwood, Jones, 1823).

205 *"There will never be a way to say definitively who did the first pushup"* Todd, "An Unexpected Pleasure."

206 *Ku appeared on the cover of a youth magazine* New York Times *Upfront*, January 31, 2022, https://upfront.scholastic.com/issues/2021-22/013122.html.

Epilogue

210 *"no / semaphoric waving, no / Rockette Rockette-flanked"* Sharon Olds, "Legs Ode," *Prairie Schooner* 86 (Summer 2012).

210 *"when someone expends the least amount of motion"* Anton Chekhov, *Anton Chekhov's Life and Thought: Selected Letters and Commentary* (Evanston, IL: Northwestern University Press, 1997).

211 *"Our bodies prime our metaphors"* James Geary, *I Is an Other* (New York: Harper Perennial, 2012).